Mismatch and Noise in Modern IC Processes

Mismatch and Noise in Modern IC Processes
Andrew Marshall

ISBN: 978-3-031-79790-3   print

ISBN: 978-3-031-79791-0   ebook

DOI: 10.1007/978-3-031-79791-0

A Publication in the Springer series

SYNTHESIS LECTURES ON DIGITAL CIRCUITS AND SYSTEMS #19

Lecture #19

Series Editor: Mitchell Thornton, Southern Methodist University

Series ISSN

ISSN 1932-3166   print

ISSN 1932-3174   electronic

Mismatch and Noise in Modern IC Processes

Andrew Marshall

SYNTHESIS LECTURES ON DIGITAL CIRCUITS AND SYSTEMS #19

ABSTRACT

Component variability, mismatch, and various noise effects are major contributors to design limitations in most modern IC processes. "Mismatch and Noise in Modern IC Processes" examines these related effects and how they affect the building block circuits of modern integrated circuits, from the perspective of a circuit designer.

Variability usually refers to a large scale variation that can occur on a wafer to wafer and lot to lot basis, and over long distances on a wafer. This phenomenon is well understood and the effects of variability are included in most integrated circuit design with the use of corner or statistical component models. Mismatch, which is the emphasis of section I of the book, is a local level of variability that leaves the characteristics of adjacent transistors unmatched. This is of particular concern in certain analog and memory systems, but also has an effect on digital logic schemes, where uncertainty is introduced into delay times, which can reduce margins and introduce 'race' conditions. Noise is a dynamic effect that causes a local mismatch or variability that can vary during operation of a circuit, and is considered in section II. Noise can be the result of atomic effects in devices or circuit interactions, and both of these are discussed in terms of analog and digital circuitry.

KEYWORDS

semiconductors, silicon, integrated circuits, variability, noise, mismatch, analog, digital, SRAM, MuGFET, silicon on insulator, reliability

Preface

Component variability, mismatch, and various noise effects are major contributors to design limitations in most modern integrated circuit (IC) processes. In this book, we take a look at these related effects and how they affect the building block circuits of modern ICs. The term variability usually refers to a large-scale variation that can occur on a wafer-to-wafer and lot-to-lot basis and over long distances on a wafer. Mismatch, which is the emphasis of Part 1, the first seven chapters, is a local level of variability that lives the characteristics of adjacent transistors mismatched. Noise is a dynamic effect that causes a local mismatch or variability that can vary during operation of a circuit and is considered in Chapters 8–12.

Chapter 1 discusses where noise, mismatch and process meet. It also contains a brief review of process and metal oxide semiconductor (MOS) components, discussion of process variability and its affect on leakage; and review of complementary metal oxide semiconductor (CMOS) gates, ring oscillators, and delay chains.

Chapter 2 takes a closer look at mismatch and variability in digital systems with analysis of mismatch effects, race conditions, and statistical modeling and with consideration of interconnect effects.

Chapters 3 and 4 consider mismatch in analog systems including in current mirrors. Cascoding, advanced mirrors mismatch effects, and minimization in current mirrors are scrutinized. Operational amplifiers (op-amps) mismatch effects and minimization in op-amps are discussed, along with circuit-induced mismatch. Memory systems are also considered.

Chapter 5 looks at reliability-induced mismatch from negative bias temperature instability (NBTI) and hot carrier injection (HCI), which are considered in both digital and analog systems.

Chapter 6 discusses nonconventional processes and circuits, in particular, silicon on insulator (SOI) and their impact on variability and mismatch.

Chapter 7 looks in detail at mismatch correction circuits and methods, including body bias, alternating current (AC) op-amps, test-and-fuse for leakage, power reduction, process control, and yield improvement.

Chapter 8 begins the discussion of noise at the component level. System noise, temperature effects, soft error rate (SER), jitter, and noise in digital systems are considered. Gate count impact on noise is also covered.

Chapter 9 considers the impact on noise on digital systems, while Chapter 10 studies noise effects in analog systems.

Chapter 11 takes a look at circuit and component design to minimize noise effects.

Chapter 12 concludes the book with a look at noise in nonconventional processes [SOI, fin-shaped field effect transistor (finFET), and multigate FET (MuGFET)].

Contents

PART I

Mismatch

CHAPTER 1

Introduction

1.1 SEMICONDUCTOR MATERIALS

Silicon is by far the most popular semiconductor for today's IC technologies; it is cheap, plentiful, and relatively easy to work with. The most versatile silicon IC process is the CMOS process, where field effect transistors (FETs) are created in the silicon by diffusion or implantation of various doping impurities. An n-channel MOS (NMOS) is constructed with source and drain diffusions, a gate region, and a substrate tie (Figure 1.1). MOS is misnomer; it originally stood for the way the transistor was constructed. Early designs used a metal gate, yet currently, MOS semiconductor structures use a polysilicon gate (self alignment). However, there are moves back toward metal-type gate in some leading edge technologies.

1.2 NMOS OPERATION

The name NMOS stands for n-channel MOS device. N-type diffusion, often phosphorous, is used to create the n-type source and drain regions (nsd). The polysilicon gate is laid down before the source and drain and masks the nsd diffusion, making the gate "self-aligned" to the source and drain for improved efficiency. There is a gate dielectric (often a grown oxide but high K dielectrics are becoming more available) between the silicon and polysilicon gate. The area under the gate is called the channel and generally is lightly doped to adjust the threshold voltage of the component within

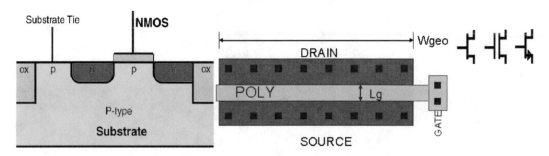

FIGURE 1.1: From *left* to *right* are simplified cross-section of an NMOS, plan view of NMOS, and symbols used for NMOS devices.

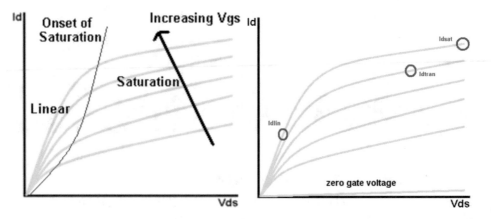

FIGURE 1.2: NMOS characteristics, showing linear, saturation, and onset of saturation regions. These correspond to specific operating states in the MOS device. The *right-hand curve* also shows the "off-state" curve.

the required range—this is the V_t adjust. Transistor characteristics are shown in Figure 1.2. The electrical characteristics of the transistor can be split into two regions, the "linear" and "saturation," with a third region, between these two, called "onset of saturation" (Figure 1.3).

1.2.1   Linear Region
Occurs when $V_{gs} > V_{th}$ and $V_{ds} < (V_{gs} - V_{th})$.

The transistor is on, and a more or less linear channel is created which allows current to flow between the drain and source (because the potential difference between the gate and the channel is approximately the same along the whole channel). The FET acts like a resistor, controlled by the gate voltage relative to the source/drain voltages. The current from drain to source is modeled as:

$$I_d = \mu_n C_{ox}(W/L)[(V_{gs} - V_{th})V_{ds} - (V_{ds}^2/2)]$$

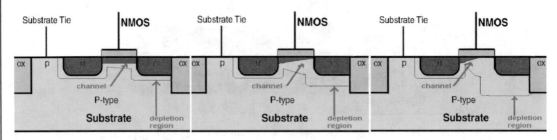

FIGURE 1.3: Linear, onset of saturation, and saturation operating conditions of a typical NMOS.

where μ_n is electron mobility, W is the gate width, L is the gate length, and C_{ox} is the gate oxide capacitance per unit area. V_{ds} is the potential between drain and source, V_{gs} is the gate-source potential, and V_{th} is the MOS device threshold voltage, the point where it begins to turn on.

When V_{ds} is much smaller than $V_{gs}-V_{th}$, the equation is simplified to:

$$I_d = \text{constant} \times V_{ds} \ (\text{Ohm's law})$$

1.2.2   Saturation Region

Occurs when $V_{gs} \geq V_{th}$ and $V_{ds} > V_{gs} - V_{th}$.

In this mode, the drain voltage is higher than or equal to the gate voltage, part of the channel is off. The onset of the off part of the channel is pinch-off. The drain current then becomes relatively independent of the drain voltage, and the current is controlled by the gate–source voltage:

$$I_d = (\mu_n C_{ox}/2)(W/L)(V_{gs}-V_{th})^2$$

When $V_{gs} \gg V_{th}$, the equation is simplified to:

$$I_d = \text{constant} \times V_{supply}^2 \ (\text{i.e., a constant current})$$

1.2.3   Channel Modulation

In reality, channel modulation adds a nonlinear trend, as there is a resistance associated with the channel, which becomes smaller the higher the drain voltage applied. As voltage gets higher, the channel gets further reduced in length and so does resistance. As resistance goes down, current continues to increase.

Output resistance approximates to:

$$R_o = (V_{ds}-V_a)/I_d$$

where V_a is an empirical voltage sometimes termed the Early voltage, after a similar effect in bipolar devices, and is defined by the voltage obtained when the saturation V–I curve is extrapolated to the X-axis (Figure 1.4).

1.2.4   Subthreshold Mode (Cutoff)

When $V_{gs} < V_{th}$, the transistor is off, and there is no channel conduction between drain and source (Figure 1.2). Ideally, there should be no current flow, except there is some due to the Boltzmann distribution of electron energies: that is, for NMOS devices, some electrons in the source have enough energy to enter the channel and flow to the drain, resulting in a subthreshold current that

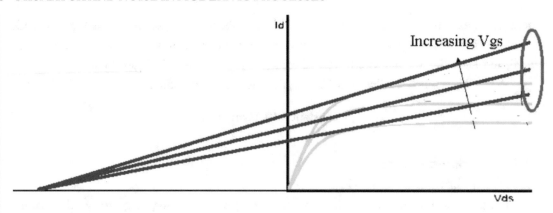

FIGURE 1.4: Extrapolation of the saturation curve back to the X-axis to determine the V_a voltage.

is an exponential function of the gate–source voltage. Thus, even at $V_{gs}=0$, there is a weak inversion current. Inversion current can be reduced further by putting a negative voltage on the gate.

1.3   PMOS DESIGN

PMOS stands for p-channel MOS device. These are similar to NMOS, except that a deep n-type diffusion (Nwell) replaces substrate as the body region and p-type source–drain (psd) replaces nsd. PMOS devices have lower drive currents per micron than NMOS in most process nodes as carriers are holes, rather than electrons, which have reduced mobility. PMOS cross-section is shown in Figure 1.5.

1.4   VARIABILITY AND MISMATCH

The point where I-drive (I_{dsat}, or any other chosen operating point) is measured gives us a single data point of operation. Figure 1.2 shows the I_{dsat}, I_{dlin} and I_{dtran} points on a standard I–V curve. In real life, we can do this check on many devices and get slightly different results for each device. We

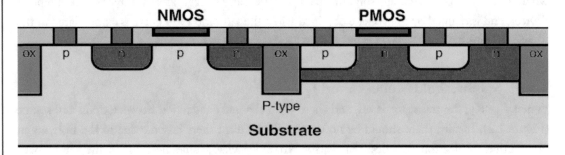

FIGURE 1.5: Simplified NMOS and PMOS cross-sections, showing how a PMOS differs from an NMOS.

have looked at nominal drive current with process node as well as strong and weak models. Strong and weak models, sometimes called fast and slow, (and those in between) result in a continuum of possible values, generally with a close to Gaussian distribution (Figure 1.6). Most circuits are designed to the weak and strong limits, which are approximately the ±3 sigma performance points. Some designs can use just the so-called corner models of weak, nominal and strong. Others require design using statistical models.

In reality, the design space or window must include limits on leakage and operating current as well as performance over a range of voltages and temperatures. The statistical models account for global variation and are applied globally across every component in the circuit. Mismatch, on the other hand, is a local phenomenon, which affects every component in the circuit individually. Combining global variability and local mismatch results in a model range that encompasses both effects. Let us assume the global distribution lies at "nominal" process corner—device-to-device variation (local variation) lies around that nominal point—same with a global distribution that lies at weak or strong, local distribution will lie around that point. Figure 1.7 shows global variations and local variations centered on weak, nominal, and strong corners (−3, 0, and +3 sigma points, respectively).

1.4.1 How Do They Happen?

Variability and mismatch are the result of anything that can be different from device to device, or anything that is made to operate in a different way due to circuit design. Systematic variability is variability that is always the same—often caused by circuit design or repeatable differences in layout, for example:

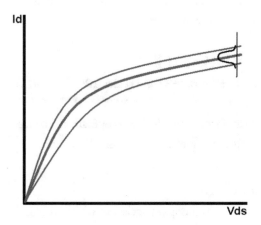

FIGURE 1.6: Demonstration of the difference between weak, nominal, and strong models (lower, middle, and upper curves, respectively), and the statistical representation obtained at the I_{dsat} point using statistical models.

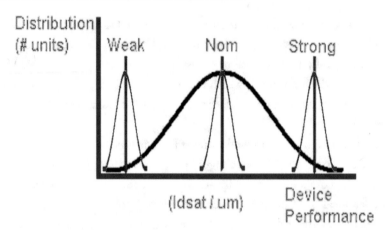

FIGURE 1.7: Demonstration of mismatch distributions superimposed on the variability distribution.

1. A transistor placed deep in an array of transistors often has a different gate length, and hence electrical characteristics, than a similar transistor at the edge of an array (especially in digital and memory circuits);
2. A mismatched circuit can result in uncertainty within the circuit to where a balance point is between matched devices (especially in op-amps).

Random variability is mismatch that can change within a device and results in variability within circuits; for example, a matched pair of transistors within an operational amplifier can lead to a circuit-to-circuit variation in the input offset. Mismatch can occur in digital circuits resulting in race conditions, or in memory circuits resulting in data loss. A transient form of mismatch can occur in certain SOI processes, where the switching history can affect delay time or offset.

1.5   CLASSES OF VARIABILITY

Physical—Device and interconnect—time scales 10–100 years (time zero effects, unlikely to change).

Functional—Changes in characteristics due to operation (device aging) (days to years).

Environmental—Changes in supply, temperature, body effect (10E-09 to 10s of seconds).

1.5.1   Random Dopant Fluctuation

Mismatch can be the result of random dopant variation within the device. A typical channel region contains just a few hundred dopant atoms, and the number of dopant atoms scales at approximately as $L_{\text{eff}}^{1.5}$. As processes shrink, therefore, mismatch due to dopant fluctuation increases. Furthermore, the location of any atom is important, and the delta V_t increases as delta N decreases.

1.5.2 Line Edge Roughness

Line edge roughness or line edge variation is a result of variation in source–drain doping, nonuni-
formities in photoresist, and polysilicon randomness; it affects the gate width and length.

1.5.3 Variability and Mismatch

This can be split into global, where wafer-to-wafer variation dominates, or batch to batch, or even
wafer fab to wafer fab. There is also within-wafer and within-reticle variation (the block of silicon
that is stepped across the wafer usually as a block of several chips that is still considered 'global'
providing it does not affect the matched circuitry within a chip). Mismatch is local variation between
adjacent devices and may be physical process-induced, circuit-induced, or environment-induced.

1.5.4 Temperature-Induced Variability/Mismatch

The effects of change in temperature on variability are evident from Figure 1.8. This is also a voltage-
dependent variability, in that a change of temperature has a much higher effect on I_{dsat} at higher
voltages and lower voltages than at the nominal conditions. This can be a global effect (if the whole

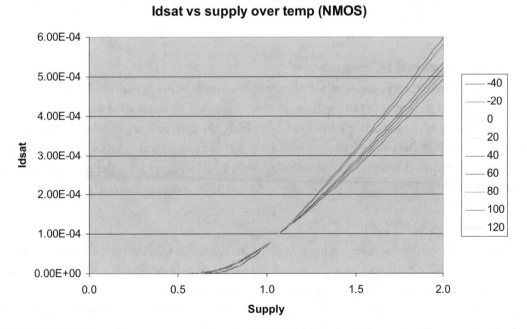

FIGURE 1.8: Effect of temperature on I_{dsat} of an NMOS. Note the difference in effect over supply, due
to the conflicting physical mechanisms which dominate at different voltages.

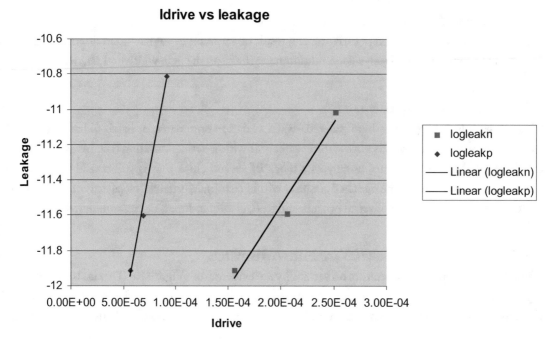

FIGURE 1.9: Leakage effects as a function of I-drive.

chip is heated) or a local effect (if some local on-chip source of heat affects one of a matched pair of devices than another).

1.5.5   Leakages and I-Drives

I_{dsat} performance of the transistors vs. log of leakage has a more or less linear correlation. As a result, a small change in performance can significantly affect leakage. This is not good when leakage power has to be considered. In the example in Figure 1.9, a PMOS I-drive increases 50% for a 10× increase in leakage, and NMOS I-drive increases 60% for a 10× increase in leakage. If variability allows a variation of 50% in I-drive, the leakage varies by about 10×. In Figure 1.10, the same is shown for mismatch, where an 8% change in performance causes a 35% change in leakage.

1.6   CMOS GATES

Static CMOS is the dominant logic gate technology in digital ICs. CMOS logic uses a combination of PMOS and NMOS devices to implement logic gates and other digital circuitry in computers, telecommunications, etc. NMOS devices generally used as pull-down devices and PMOS generally act as pull-up devices. Figure 1.11 shows circuit diagrams of inverter, NAND, and NOR CMOS logic gates.

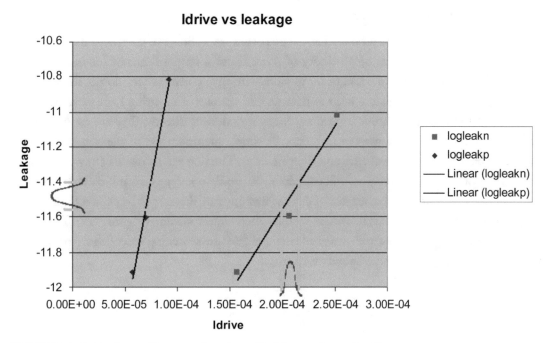

FIGURE 1.10: Leakage effects as a function of I-drive—mismatch effects.

The simplest gate is the inverter. Its performance is defined by the component and interconnect parasitics of the turn-on switching devices and is directly a function of I-drive and capacitance. If variability and mismatch affect transistors, they therefore equally affect more complex circuits. To test CMOS gates, it is normal to connect groups of inverters or other gates as chains or rings in order to be able to analyze the average performance of such gates.

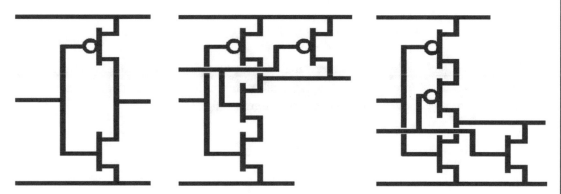

FIGURE 1.11: From *left* to *right* are schematics of typical CMOS inverter (sometimes called a NOT gate), NAND, and NOR gates.

1.7   RING OSCILLATORS AND GATE CHAINS

Ring oscillators and chains are used to check logic performance. Some of the things they can check are optimal layout, the effects of proximity of other features on performance (dense and sparse layout). Rings are more versatile and easier to work with than chains as they do not need external switching signals. Ring oscillators need to have an odd number of inverting logic gates, and since frequency is too high to take out to external measuring devices directly, they are connected to a divider, which can then measure the frequency. A simple multiplication of the output frequency then leads back to the ring frequency and the gate delay. The ring can be made of any type of gate, although most often any given gate will contain all or mostly one type or combination of types of gate. Similarly fan-out (i.e., the number of devices that any gate drives) can be chosen for any given ring oscillator, as can other characteristics such as interconnect loading (Figure 1.12). The things that can be determined from a well-designed ring oscillator are switching speed/logic performance, leakage, power dissipation, and reliability verification.

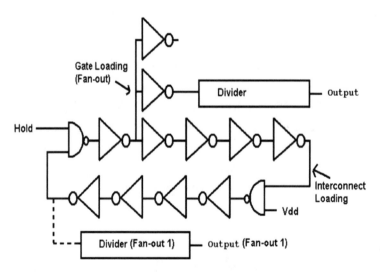

FIGURE 1.12: Representation of a ring oscillator, showing how placements of interconnect, loading, and differing gates can be incorporated into a ring oscillator (normally a ring oscillator contains only one device type or fan-out). The 'hold' function is added to check off-state leakages.

C H A P T E R 2

Variability and Mismatch in Digital Systems

2.1 EFFECT OF INTERCONNECT

Interconnect capacitances and resistances are significant. They can cause frequency of test configurations and circuits to drop by 50% or more. As a result, any variability or mismatch that occurs in the interconnect has a direct impact on the overall circuit performance. Mostly, interconnect can be simulated with resistor-capacitor (RC) networks (Figure 2.1). Complexity of the resulting circuits, however, is formidable, and as a result, analysis of circuits with full parasitic extraction is limited to relatively small circuits or subcircuits that have suspect or unknown performance. Logic is often not analyzed as a spice subcircuit at all, but with Verilog or some other digital simulator—this is because of the huge quantities of logic in most ICs. A lumped parasitic may be added into the digital simulator without a big increase in simulation time. Analog circuitry, however, often requires the full parasitic extraction, combined with an analog simulator.

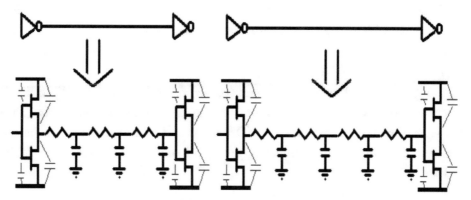

FIGURE 2.1: Interconnect can generally be simulated with RC arrays, which are extracted from the layout by special CAD tools.

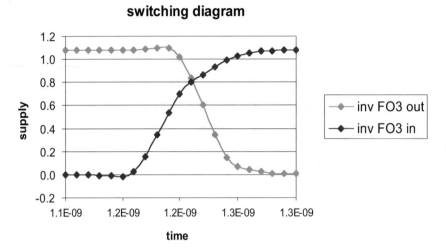

FIGURE 2.2: Switching of input and output. The slowing of the switching of the input is a result of Miller capacitance during output switching.

2.2   RING OSCILLATOR SWITCHING CHARACTERISTICS

The input to each gate stage switches (from 0 to 1 or from 1 to 0), followed at some time later by the output (switching the other way), is dependent upon transistor sizes, loading, temperature, and supply. At any given moment during the switching, there is a gate and drain voltage for the PMOS and NMOS. Switching performance (delay) is determined by every point on the curve. Capacitance is not a constant here, being defined by all the fixed capacitances of the dielectric capacitances (interconnect), the junction capacitances (drain to body), and Miller multiplication of capacitance across the gate. Figure 2.2 shows the switching of the input and output. The slowing of the switching of the input is a result of Miller capacitance during output switching.

2.3   DIGITAL CORRELATION

Referring to Figure 2.3 [1,2], consider the input (yellow) rising from V_{ss} (time 'A') to the time the output (blue) just begins to turn on—this is the delay through the gate. The instant that the output begins to switch is the same point that the PMOS current and NMOS current are in balance. The input voltage at which that occurs is defined by the difference between the V_{tsat} of the NMOS and the V_{tlin} of the PMOS, or:

$$\text{Voltage threshold rise } (V_{tr}) = V_{tsatn} + 0.5(V_{dd} - V_{tsatn} - V_{tlinp}),$$

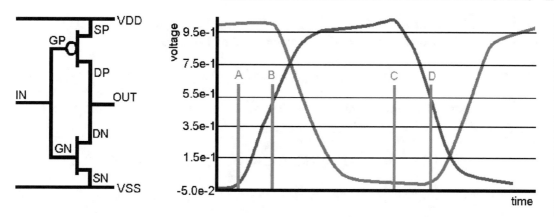

FIGURE 2.3: Input and output waveforms of an inverter in a ring oscillator, showing the time A–B (C–D), where the input is rising (falling) ahead of the output that is beginning to switch.

and the falling input voltage (delay C–D) is:

$$\text{Voltage threshold fall } (V_f) = V_{tsatp} + 0.5(V_{dd} - V_{tsatp} - V_{tlinn})$$

when measured as a voltage swing from supply. Or the above equations can be expressed as:

$$V_r = (V_{dd} - V_{tlinp} + V_{tsatn})/2 \text{ and}$$

$$V_f = (V_{dd} - V_{tlinn} + V_{tsatp})/2.$$

The time from start of input rise and output fall are:

$$t_r = V_r c/pt \text{ and}$$

$$t_f = V_f c/nt,$$

where c is the capacitance and is approximately the same for rising and falling slopes, nt for I_{dtrann} and pt for I_{dtranp} are average values of the I_d of the NMOS and PMOS during switching.

Thus, total delay (one rise and one fall) is:

$$\text{Delay} = [c(V_{dd} - V_{tlinp} + V_{tsatn})/2pt] + [c(V_{dd} - V_{tlinn} + V_{tsatp})/2nt],$$

and the frequency is:

$$\text{Frequency} = 2ptnt/\{c[nt(V_{dd} - V_{tlinp} + V_{tsatn}) + pt(V_{dd} - V_{tlinn} + V_{tsatp})]\}$$

2.3.1   Defining I_{dtran} Voltage

From Figure 2.4, the output start switch to next output start switching is:

$$V_{gs}\,min\sim(V_{dd}-V_{tlinp}+V_{tsatn})/2,\ V_{gs}\,max\sim0.95V_{dd}.$$

The maximum V_{ds} during this time is V_{dd}, and the min V_{ds} is approximately $(V_{dd}-V_{tlinn}+V_{tsatp})/2$. The first order average value of $V_{ds}=V_{gs}$ is:

$$=[V_{dd}+(V_{dd}-V_{tlin}+V_{tsat})/2]/2=0.75V_{dd}+0.25(V_{tsat}-V_{tlin})\tilde{}\,0.75V_{dd}$$

We have shown that:

$$\text{Delay}=[c(V_{dd}-V_{tlinp}+V_{tsatn})/2pt]+[c(V_{dd}-V_{tlinn}+V_{tsatp})/nt]$$

Let us look at variability in digital systems—assume that I_{dsat} varies, first in a global manner. For the purpose of this exercise, let us say I_{dsat} is the same as I_{dtran} (nt, pt), and all other variables stay the same, then:

$$\text{Delay fall}=K_aI_{dsat}(N)^{-1}\text{ and}$$

$$\text{Delay rise}=K_bI_{dsat}(P)^{-1}.$$

A 1% increase in I_{dsat} (equivalent to a 1% local mismatch increase) reduces delay by approximately 1%.

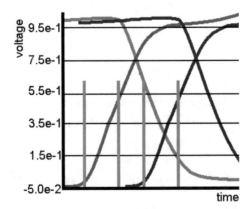

FIGURE 2.4: Switching response to determine average input and output conditions during performance defining part of switching response.

Combining using sum of squares will get:

$$\text{Average delta delay (ADD)} = [(K_a I_{dsat} N^{-1})^2 + (K_b I_{dsat} P^{-1})^2]^{0.5}$$

If $K_a = K_b$ and $I_{dsat}N = I_{dsat}P$:

$$\text{ADD} = \text{sqrt}(2)K/I_{dsat}.$$

Thus, if we have an inverter made of two transistors of equal variability, then our delay variability will be 1.4× the individual transistor variability (i.e., if the variations are 100% independent of each other). For example, channel doping is completely independent.

However, if the variability of the transistors is identical, such as oxide thickness, we cannot combine the variations using sum of squares, and variability becomes 2× the individual transistor variability.

2.4   SIMULATION OF STATISTICAL AND MISMATCH EFFECTS

As shown in Figure 1.7, the development of a statistical or mismatch model is generated as follows. First, a nominal model is created. From that corner, weak and strong models are created. These are like nominal, but a few select parameters are modified to weaken or strengthen the model. A statistical model is created by applying globally (to all devices in the circuit) the same offset—these are the standard deviation (snd) parameters. Mismatch is an additional parameter that can be applied to individual transistors, designated "mm" parameters.

2.4.1   Variability and Mismatch Parameters

Statistical analysis methodology applies weighing to generate process performance under different physical conditions, weak and strong models are specific process corners. For MOS devices, several variables can be used for modifying the component characteristics:

- L_r (gate length)
- W_r (gate width)
- C_{ox} (gate oxide capacitance)
- N_{vfb} (NMOS flat-band voltage)
- P_{vfb} (PMOS flat-band voltage)

Flat-band voltage (vfb) in MOS devices refers to a voltage at in the band diagram where the energy bands of the semiconductor are flat.

Additional model variables may include channel doping variation and mobility variation.

Mismatch uses a subset of these variables, which can vary by process but usually includes a gate length and/or width variable and a channel doping variable.

Plotting statistical analysis of a delay chain against supply (Figure 2.5) shows that the delay spread increases with lower voltage, indicating a high dependence on V_t variation. In this graph showing an older technology node, nominal, weak, and strong models are shown to be in good agreement with mean, +3 sigma, and −3 sigma corners at high voltages, but they diverge at lower voltages.

Another important aspect of any circuit design is leakage current. The spread here is shown in Figure 2.6, where at the higher voltages leakage is higher, as expected. However, since this is the high-voltage corner of the process, the correlation between statistical and conventional 'corner' models is reasonable.

Correlation is improved at low voltages with the introduction of the concept of current correlation coefficients (Figure 2.7). By modifying individual coefficients to match with the trends of the corner models over supply voltage, temperature, or any other variable of interest, the models are made to match in performance and leakage over the same variable.

Where do we need statistical models? Where variation of every transistor/component/interconnect on a chip varies by same amount—ideal for absolute timing, absolute performance and where local mismatch can be averaged out.

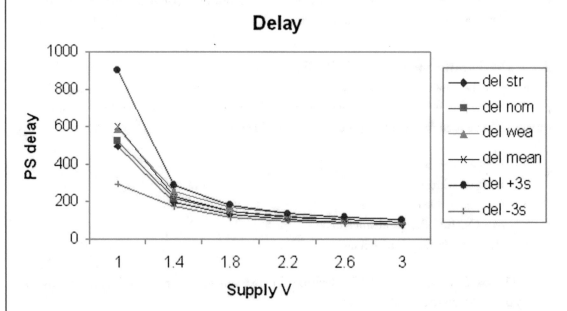

FIGURE 2.5: Process delay spread with voltage for nominal, weak, and strong models and mean, +3 sigma, and −3 sigma corners.

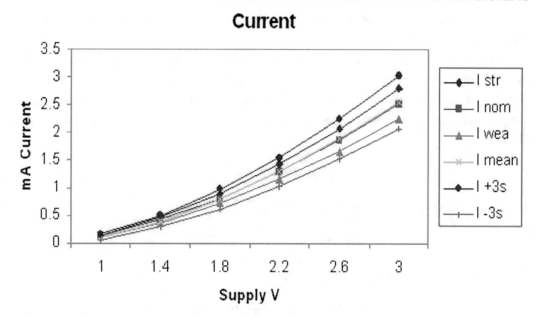

FIGURE 2.6: Circuit leakage spread with voltage, for nominal, weak, and strong models and mean, +3 sigma, and −3 sigma corners.

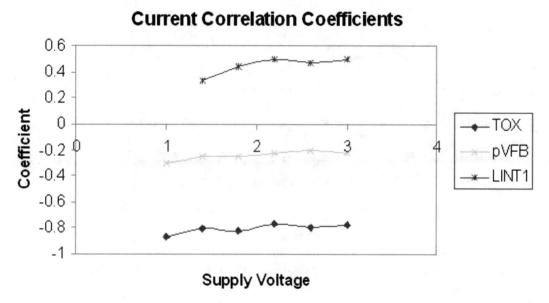

FIGURE 2.7: Example of the way correlation coefficients of the model can be linearized over supply voltage to give improved matching between corner and statistical models (after linearizing).

Where do we need mismatch analysis? Where local variation is important—SRAM cell, race conditions in short paths, amplifiers.

But, what is the definition of where we can neglect local mismatch in real circuit design? Since mismatch is completely random as the number of components or logic gates increases, the relative effect of the mismatch is reduced (delay increase is linear, mismatch is the square root function). Assuming that we have a gate delay D, with a mismatch delay delta of d, the gate delay will be:

$$\text{Gate delay} = D - D' \pm d.$$

For circuits with two gates, the average delay is:

$$\text{Average delay} = D \pm \text{sqrt}(2)d/2,$$

while for cicuits with N gates, the average delay is:

$$\text{Average delay} = D \pm \text{sqrt}(N)\,d/N.$$

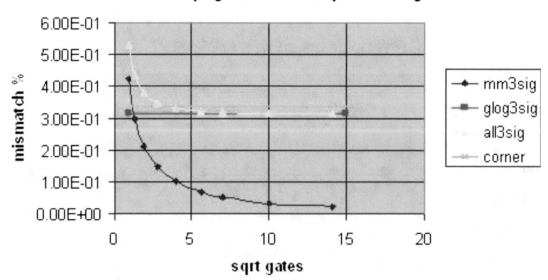

FIGURE 2.8: Mismatch, global, corner, and combined mismatch as a function of gate length of a logic chain.

If d is 10% of D and we need the effect of d to be < 1%, then we would need sqrt(N)=10 or N=100. If d is 10% of D and we need the effect of d to be < 5%, then we would need sqrt(N)=2 or N=4.

Considering Figure 2.8, we see how mismatch averaged over the number of gates in the chain quickly reduces to the point where overall mismatch (local and global variability) reduces below that of the corner model, thus permitting simpler corner model circuit analysis for all, but short chains and specific other circuits.

2.5   VERIFYING RANDOMNESS

One way to check for mismatch is to use an addressable FET array (Figure 2.9). Here, it is possible to place a large array of transistors in an array and select the one(s) of interest, with a switching matrix. This is good for most on-state measurements, but does not work well for leakage measurements, as there are several leakage paths.

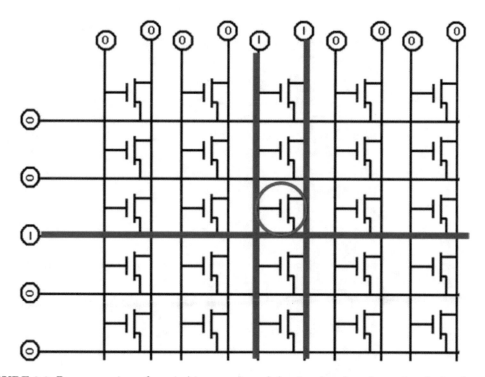

FIGURE 2.9: Representation of a switching matrix and the signal path to determine device characteristics.

CHAPTER 3

Variability and Mismatch in Analog Systems I

3.1   CURRENT MIRRORS

NMOS and PMOS current mirrors are arguably the simplest analog circuit and are widely used in integrated circuits. A single current source acts as an input and mirrors multiple identical currents. Matching is typically about 1% and can be optimized with component and circuitry design and layout techniques. A simple current mirror is shown in Figure 3.1.

The current mirror concept is simple: If all the voltages on an output device are the same as on an identical input component, then all the currents flowing through the devices are also identical. Then:

Input current = Output current

So what is reality?

- What effect does global variation have on a current mirror?
- What effect does local mismatch have on the current mirror?

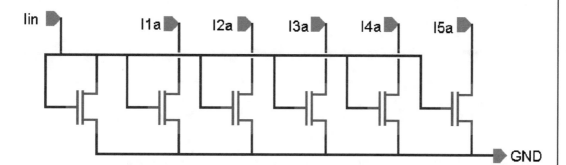

FIGURE 3.1: Basic current mirror with an input (*left-hand component*) and five outputs. The circuit is used to 'mirror' an input current, allowing multiple matched currents to be used in different circuits or areas of a circuit.

Well, it depends, but let us start by assuming that we are using minimum gate length and width components, similar to those used for the best, fastest, lowest power digital circuits.

3.2 GLOBAL VARIATION

Figure 3.2 shows how a current mirror varies as a function of global variability (where all components change the same). Here, the voltage on the input (reference gate and drain) shifts with model strength, but output current at the same output drain voltage is constant. Away from this singularity, significant variation can occur even when there is no local mismatch.

An analysis of Figure 3.3 demonstrates that when mismatch is applied to the circuit in Figure 3.1, there is significant difference in the current. Some of the outputs match fairly well at the balance point, but others are off by up to 80%. This begins to highlight a problem with the use of minimum size transistors, which are so ideal for digital purposes but are not so good for analog purposes. Note that this is a snapshot. The next time a mismatch analysis is simulated, since this is a statistical process, we might expect different results, which could be better or worse.

To improve this, we can reduce variability of mismatch parameters Lint (m), width (m), NCH (N-type channel doping). In the case of NCH, the channel doping, this is a process effect, which is difficult to change except when a process is being developed. However, an increase in width and length of transistors is readily achieved. An increase in length might initially appear to improve

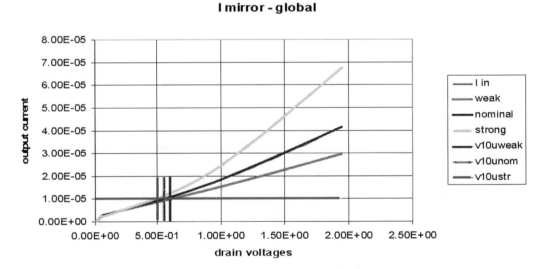

FIGURE 3.2: While the balance point remains basically the same, stronger or weaker models result in a change in the output current mirroring under conditions away from the point where all voltages are the same.

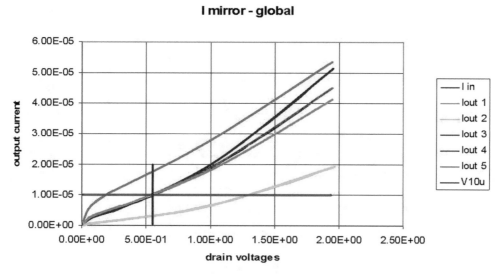

FIGURE 3.3: Plot of the output currents of Figure 3.1, showing significant potential for mismatch-induced error even at the normal matched current point.

matching, but this is happening as a result of saturation of the transistor, so the matching improvement cannot be relied upon if mirror currents or temperature effects are introduced (Figure 3.4).

Increase in width will improve the mismatch somewhat, but can lead to higher mismatch outside the matching voltage point (Figure 3.5). This is because voltage V_t is reduced (the input device is wider, thus, the voltage needed across its gate/drain is lower to maintain the same current). This causes more mismatch as $V_{gate}-V_t$ is reduced.

One thing we might do is try increasing both width and length of the transistors. This is advantageous as mismatch is reduced, and it can be designed such that the other electrical characteristics, such as voltage V_{gate}, can be raised out of the region where mismatch becomes serious. The resultant design is simulated as shown in Figure 3.6. The resultant output current over the operating range of 0.5–2 V gives an output current range of over 2:1, thus, it is better but still not as good as we would like.

Hence, theory matches results, and we can increase W and L to get improvement in current mirror matching. This is termed component optimization, but in terms of actual silicon, we have a need to do several additional things to optimize the mirror. We can add dummy devices surrounding the mirror components, which improves matching as the end devices are no longer edge devices, where the photolithography tends to pattern the gate differently, resulting in devices with somewhat different gate lengths (Figure 3.7). Transistors should also all be aligned in the same orientation,

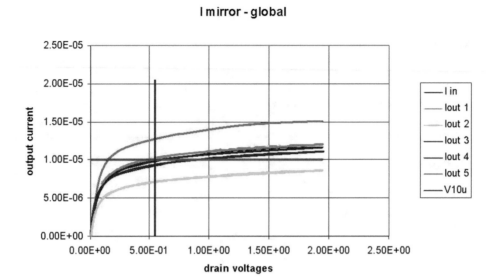

FIGURE 3.4: Graph showing matching of longer channel devices. The matching improvements seen here are not reliable because they are dependent upon the individual transistor characteristics, not the circuit.

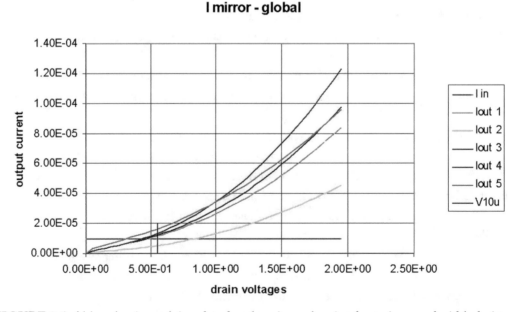

FIGURE 3.5: Although mismatch is reduced at the mismatch point due to increased width devices, at higher voltages, the matching is reduced because the lower diode voltage causes more component variability.

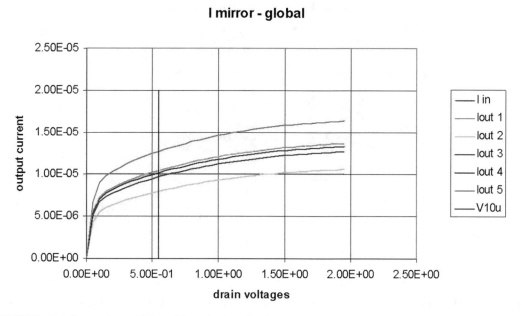

FIGURE 3.6: Increasing width and length gives further improvement to mismatch, but better matching is needed in many systems.

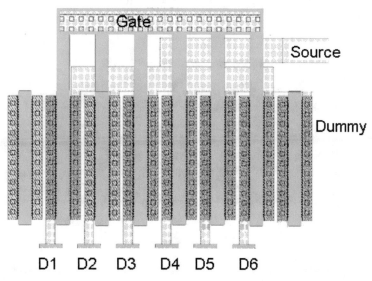

FIGURE 3.7: Layout of current mirrors, showing transistor alignment and dummy devices.

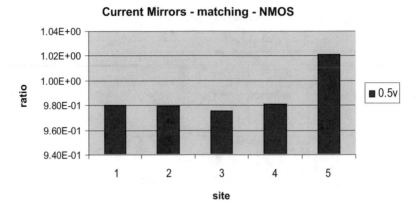

FIGURE 3.8: Representation of matching when there are an inadequate number of dummy devices around the reference device.

i.e., all the drains to the left and all the sources to the right of the component, in this example. This reduces variation due to global process variation. An additional layout optimization that should be done is resistance matching of interconnect.

With inadequate mismatch correction at layout, we can see results that look similar to Figures 3.8 and 3.9. In Figure 3.8, there is inadequate compensation for mismatch due to the edge effects, i.e., insufficient dummy devices. The result is that current for the output devices within the array are lower due to the outer devices having a lower V_t than the inner devices. Device number 5 (D5), another edge device, has a slightly higher current due to its closer matching to the input reference device.

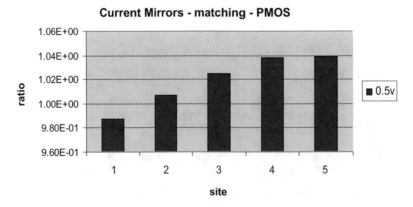

FIGURE 3.9: Representation of matching as a function of distance from the reference device.

Figure 3.9 shows the effect of reduced matching with distance from the reference device. For critical applications, it is normal to place the reference device in the middle of the array; thus, the best averaging is achieved.

3.3   CASCODE CURRENT MIRROR

We have looked at simple current mirrors and some layout techniques for optimization, but we do not have as good matching as we would like or is often required. Using circuit design techniques, can we improve performance (current mirror matching) by modifying the circuit in some way?

Fortunately, yes. Let us simplify by modifying the circuit to just one input and one output (Figure 3.10). Using what is known as cascoding techniques, we can apply a circuit that gives a simple reference voltage that is approximately equal at the input and output. In Figure 3.10, this corresponds to nodes n_C and n_D. If n_C and n_D are equal, we operate at the same voltage for the input and output of the mirror. Under these conditions, we operate near the matched current point throughout the operating range. The simple current mirror in this case gives a variability of 7.5–15 μA (Figure 3.11), but if the operating range of the output of the mirror transistor has a restricted voltage range, the variability is reduced further to 7.5–13 μA. This gives us a mirroring ratio of better than 1:2, which is within the usable range, for many requirements. Figure 3.12 shows a typical circuit configuration which achieves the voltage regulation of Figure 3.10.

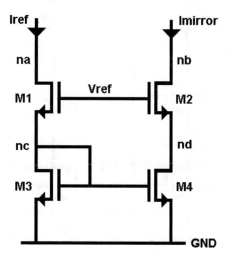

FIGURE 3.10: Circuit showing a cascoded configuration. This restricts the maximum operating range of the mirroring transistor and improves matching.

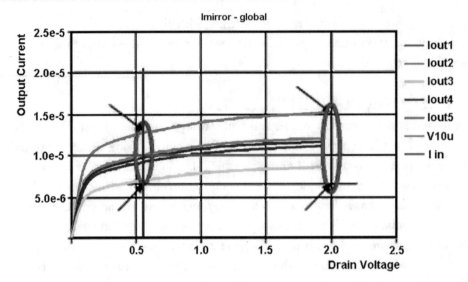

FIGURE 3.11: The output current range of a noncascoded mirror is defined by the minimum current at the lowest voltage and the maximum current at the highest operating condition. With a cascoded mirror, the maximum current is defined by the current of the source voltage of the output cascode.

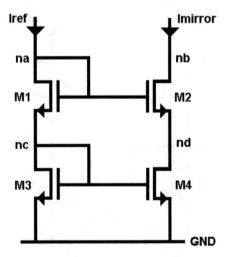

FIGURE 3.12: Typical simple cascode circuit, using the input source to provide the n_C and n_D reference voltages.

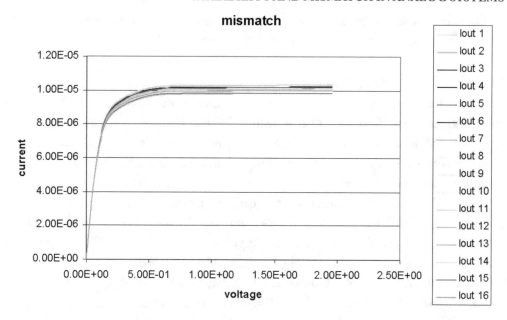

FIGURE 3.13: Optimized simple cascode current mirror, simulated with mismatch models. This indicates approximately ±4% accuracy can be obtained with good layout, even with mismatch.

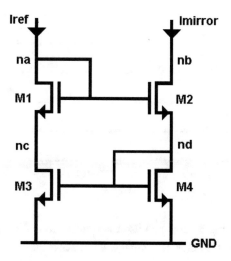

FIGURE 3.14: Super Wilson current mirror. This design has some matching advantages over the cascoded mirror for precision matching.

Optimization of this simple cascode circuit and application of mismatch simulation leads to an output current range of 9.64–10.4 μA or 1:1.079 (or ±4%) (see Figure 3.13).

3.4   WILSON CURRENT MIRROR

On the quest for more precision, we appear to be closing on the limit with the traditional and cascoded current mirror. Can we do better with a different design? One other standard mirror structure is the Wilson current mirror (Figure 3.14). The Wilson mirror by itself has some of the disadvantages of an un-cascoded conventional mirror, but, using a 'super Wilson' mirror, we get better matching—what can we expect?

A super Wilson mirror is supposed to give better matching; with the optimized conventional scheme, MM is 9.64–10.4 μA (1:1.079 or ±4%), but with the super Wilson, MM is 9.08–10.23 μA (1:1.13 or ±6.5%). Looking closely at the output waveform (Figure 3.15), we notice that at the low-end voltages, the current is dropping off faster. If we look instead at a low corner point of 0.7 V instead of 0.5 V, the optimized conventional scheme, MM is 9.78–10.4 μA (1:1.063), whereas the super Wilson, MM is 9.67–10.23 μA (1:1.058). Thus, under slightly restricted operating conditions, the Wilson mirror is indeed an improvement over conventional designs.

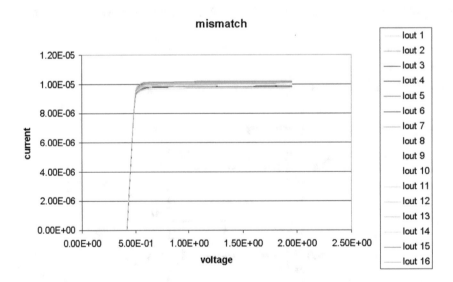

FIGURE 3.15: Matching using a supper Wilson current mirror. At 0.5 V output, the currents have already begun a steep drop; this is below the voltage where reasonable matching can be expected.

CHAPTER 4

Variability and Mismatch in Analog Systems II

4.1   OPERATIONAL AMPLIFIER

The basic CMOS operational amplifier (op-amp), as shown in Figure 4.1, requires a matched input pair, the tail current is balanced with the output current, the PMOS mirror acts as a balance for transistor M_5 pull-up. In addition, there is an RC high-frequency filtering to improve stability.

   The matched current sources are generally generated from current mirrors (Figure 4.2); usually, a simple current mirror is used. This simple circuit already has some circuit-matching issues, even under 'balanced' conditions; M_7 and M_8 drains are not likely to be the same voltage, and with a simple current mirror, currents will therefore not be the same. Currents are V_{in}-dependent; hence, if V_{in} is close to V_{dd}, better balance is achieved. Likewise, M_3, M_4, and M_5 drain voltages are not likely to be equally balanced.

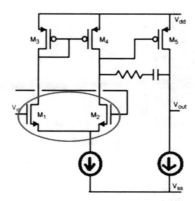

FIGURE 4.1: Basic op-amp, showing the input pair. The input pair is the part of the circuit most sensitive to mismatch.

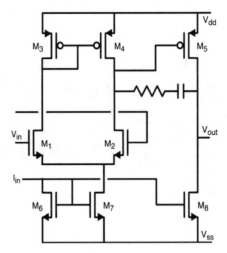

FIGURE 4.2: Op-amp, conventional with current mirror current sources.

4.1.1 Input Pair Matching

Standard practice in precision op-amps is to cross-couple the inputs; hence, the noninverting input, for example, in Figure 4.3, would be the 'a' and 'd' devices, and the inverting input would then be the 'b' and 'c' devices. The sources are all common, and the drains of 'a' and 'd' and 'b' and 'c' are common,

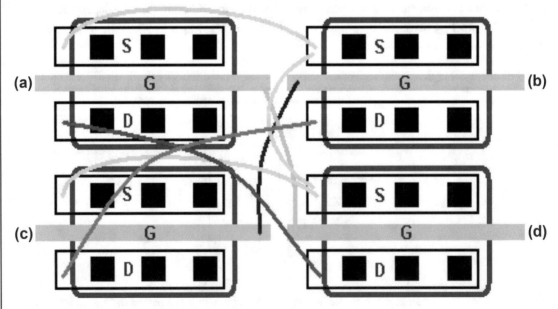

FIGURE 4.3: Representation of input pair cross-coupling, used in precision op-amp input pairs.

as well as the gates. The sources and drains are all aligned, so if any processing misalignment was to occur, all devices would be equally affected. In high-precision designs, dummy devices surround the active devices. In addition to processing mismatch, the cross-coupled design gives some immunity to environmental mismatch, in particular, thermal or stress mismatch.

4.2   IDEAL OP-AMP

For circuit design, we generally use a simplification and assume an ideal or close-to-ideal amplifier. Figure 4.4 shows that the main part in an amplifier is the dependent voltage source. This increases as a function of the voltage drop across R_{in}, amplifying the voltage difference between V^+ and V^-. The ideal op-amp has infinite open-loop gain, infinite bandwidth, zero input current, zero offset voltage, infinite slew rate, zero output impedance, and zero noise.

4.2.1   Open-Loop Gain

The amplifier's differential inputs consist of V^+ input and a V^- input, and generally the op-amp amplifies only the difference in voltage between the two. This is called the 'differential input voltage.' This configuration is generally used to build a comparator function, where a change in voltage of the input pair (say as one input voltage terminal switches from being less than the other to more than the other) causes a full rail swing on the output, say from V_{ss} to V_{dd}. In open-loop mode, any input offset is amplified. In the ideal case (Figure 4.5a):

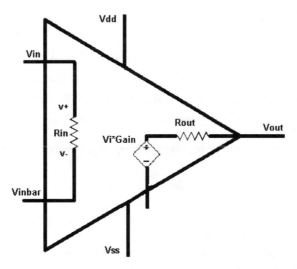

FIGURE 4.4: Op-amp block diagram—this simplification is often used for circuit design. Mismatch can be applied to the ideal op-amp circuit.

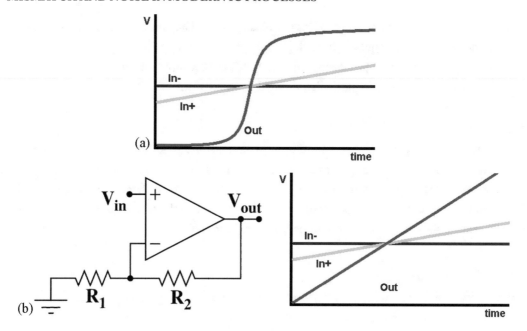

FIGURE 4.5: (a) Simple comparator waveform. When the rising input meets the voltage of the steady input, switching occurs at the output. (b) Negative feedback schematic and representation of op-amp output for a system with controlled gain such as a negative feedback system.

$$V_{out} = (V_{in}-V_{inbar})\text{gain}.$$

4.2.2   Negative Feedback Operation

By applying resistive feedback, the gain can be controlled. This configuration is generally used to build a linear amplifier, where the output voltage is a small multiple of the difference in input voltage pins. In this mode, any input offset is amplified, but only by the gain defined by the negative feedback. In the ideal case (open-loop gain=infinite), i.e., $V_{in}+=V_{in}-$, thus:

$$V_{out}=V_{in}\ (R_2+R_1)/R_1.$$

Feedback and output waveform are shown in Figure 4.5b.

4.2.3   DC Imperfections in Op-Amps

Gain of the amplifier is not infinite. Finite gain affects all systems where feedback aims for a gain that is higher than about 10% of the actual gain of the amplifier.

Output resistance becomes a factor when driving low-resistance loads and low-resistance feedback loops.

Input bias current. Not usually a problem in CMOS op-amps, but any input current, however caused, tends to generate an input offset, which is gained up through the amplifier and can result in systematic offset.

Input offset voltage. This is the voltage required across the op-amp's input terminals to drive the output voltage to zero.

Nonlinear common mode gain. Manifests itself as an input voltage dependency.

Temperature effects. All parameters change with temperature. Temperature drift of the input offset voltage is especially important.

Finite input resistance puts an upper bound on resistance of the feedback circuit. Some op-amps have circuitry to protect inputs from excessive voltage which can have a resistance to a supply voltage.

4.3   FINITE GAIN

A typical MOS op-amp may have a gain in the order of 500–50,000.

$$V_{out}/V_{in} = G_{ol}G_{res}/(G_{res}+G_{ol}),$$

where:

$$G_{res} \text{ (gain defined by feedback loop)} = (R_1 + R_2)/R_1, \text{ and}$$

$$G_{ol} \text{ (open-loop gain)} = V_{out}/(V^+ - V^-).$$

Hence, if G_{ol} is infinite and G_{res}=100, V_{out}=100V_{in}, but if G_{ol}=500 and G_{res}=100, V_{out}=83.3V_{in}, or if G_{ol}=50 and G_{res}=200, V_{out}=40V_{in}.

With an expected gain of 100 but an amplifier gain of 500, instead of infinite, the actual gain is only 83.3. Meanwhile, if the amplifier gain is only 50, but a gain defined by the resistance feedback is 200, the actual gain observed is only 40.

4.3.1   Nonzero Output Resistance

Generally, we consider the output impedance as infinite; however, it usually is not. For direct current (DC) purposes, we only need to be concerned about resistance. When M_5 is off (Figure 4.6), the

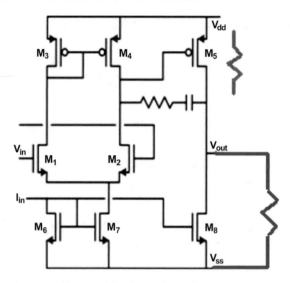

FIGURE 4.6: Output resistance effects and loading of a real op-amp.

output resistance will pull down to ground, thus no difference there, but when M_5 is on, we have a resistive divider.

Under comparator operation, the output voltage will only rise to:

$$V_{out}=V_{dd}R_{out}/(R_{out}+RM_5),$$

neglecting the current through M_8.

Under ideal conditions V_{out} is in the middle of its switching operation when V_{in} and V_{inbar} are equal. Under conditions of finite output impedance, offset will be modified. If current through R_{out}=current through M_8: then an offset is needed on M_2 (gate voltage of $M_2>M_1$) for the output to balance.

4.3.2   Input Bias Current

A small current (typically picoamperes for CMOS designs) flows into the inputs. Some low-voltage MOS gates have enough leakage that gain may be changed, but most do not; in general, op-amps use higher-voltage devices for matching reasons. High-voltage components have thicker gate oxides, hence have lower leakage. One potential area where gate leakage is important is in sample and hold circuits (Figure 4.7), where leakage means faster refresh is necessary. In addition, electrostatic discharge (ESD) protection circuitry usually added to external pins can leak, causing errors (Figure 4.7).

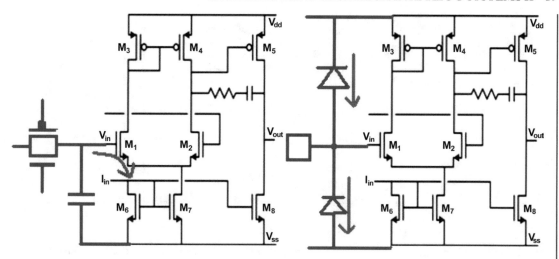

FIGURE 4.7: Sample and hold (*left*) is sensitive to any input leakage that may be present and ESD protection circuitry (*right*) induces leakage.

4.3.3 Input Offset Voltage

Input offset voltage is the voltage required across the op-amp's input terminals to drive the output voltage to zero. This is mismatch/variability of components viewed from the point of view of a circuit black-box. The simplest way to view this is as an equal but opposite voltage source on the input (Figure 4.8).

As show in Figure 4.9, the offset voltage, in reality, is a combination of four offsets: systematic (circuit design, V_{cd}), systematic (layout, V_{lay}), global random (global variability, V_{gl}), and local

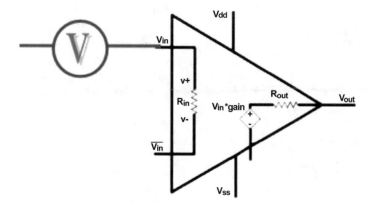

FIGURE 4.8: The simplest way to consider input offset voltage in the op-amp model is as an external voltage supply.

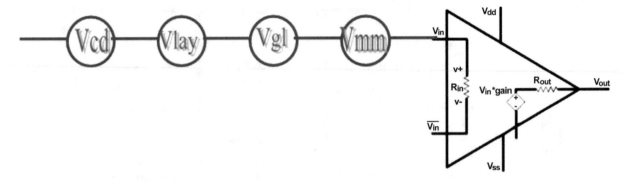

FIGURE 4.9: Overall offset voltage, showing the combination of offsets that results in the total offset.

random (local mismatch, V_{mm}). The systematic offsets can be designed outside, to a large extent. Component sizing can help with the random offsets, but these are much more difficult to design.

4.4 COMMON MODE GAIN

A perfect op-amp amplifies only the voltage difference between its two inputs, completely rejecting all voltages that are common to both. Simple op-amps such as those in Figure 4.2 are not good at rejecting the common mode voltage, as current alters through the input pair with supply, but remains unaltered through the output device. This mismatch causes a drift in output voltage with input.

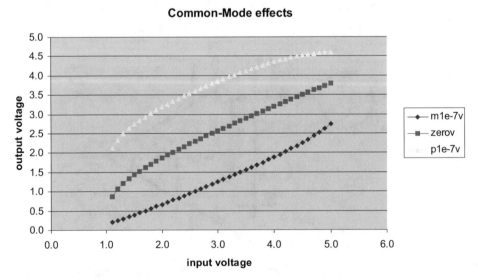

FIGURE 4.10: Effect on gain of the common mode, for an input offset of −0.1 mV, 0 V, and +0.1 mV.

This is demonstrated in a 5 V op-amp of the design of Figure 4.2, when input is ramped from 0 to 5 V, while input bar ramped either with 0 or ±0.1 mV. The perfect amplifier would show no voltage swing with input variation, but Figure 4.10 demonstrates a variation, say at an output of 2.5 V, where a 0.2-mV input offset is equivalent to an input voltage range of 1.2–4.8 V or 3.6 V.

Sanity check. As input voltage increases (Figure 4.2), the balance point where current M_3, M_4, and M_5 is equal increases. Since the balance point current is higher, there will be a stronger pull-up. As a result, V_{out} is pulled higher, as more current flows through M_5. Therefore, what we are observing does make sense.

4.5   REDUCING THE COMMON MODE EFFECT

The most significant impact on common mode offset is the quality of the current mirroring, which was dealt with in Chapter 3. We can apply the lessons learned from there to improve the current mirroring of the op-amp we are now optimizing. We might start by lengthening M_6, M_7, and M_8 (in this case, from 1 to 5 μm). The result is a much flatter curve and reduced common mode effect (Figure 4.11), although we have added a systematic offset, due to limitation of the output current, thus moving the balance point up closer to the PMOS V_t.

Long channel helps the common mode rejection, but based on previous results for current mirrors, we can expect better results from cascoding. A modified circuit (Figure 4.12) shows the

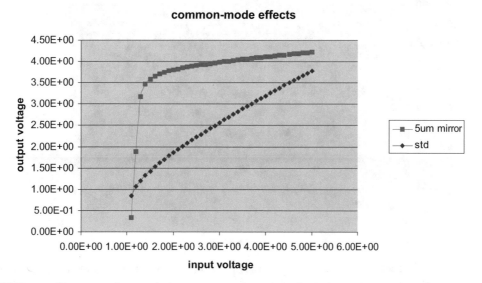

FIGURE 4.11: Variation of output balance point with a relatively short gate length (1 μm) current mirror and long channel (5 μm) current mirror.

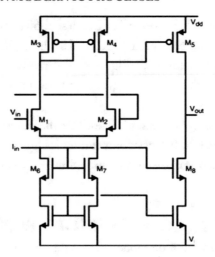

FIGURE 4.12: Addition of cascoded current mirroring to the op-amp gives an improved common mode rejection.

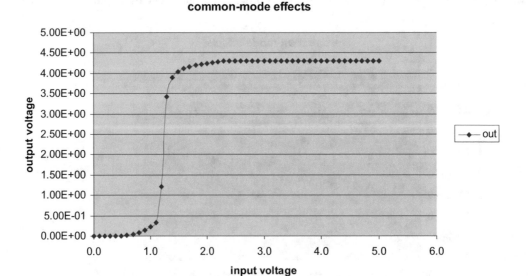

FIGURE 4.13: Cascoded current mirroring to the op-amp gives an improved common mode rejection, over about 1.5 V.

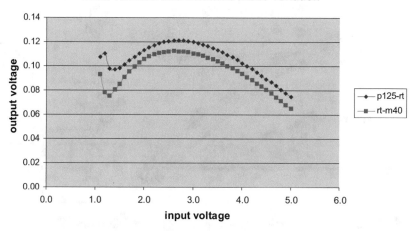

FIGURE 4.14: Difference in offset between room temperature and 125°C and −40°C as a function of input voltage, for a conventional op-amp.

simplest cascoded circuit. This would not be expected to perform well at low voltages, but should give a good rejection at higher voltages. As expected, this looks much better, once we are above the mirror operating point (Figure 4.13).

4.6   TEMPERATURE EFFECTS

Component characteristics change with temperature. Of most importance in op-amps is the variation of V_t, which will impact the input pair source voltage. In amplifiers with conventional current mirror sources, this, at any given input voltage, will affect the tail current. In effect, as temperature rises and V_t reduces, this will increase the input pair voltage, acting as a common mode offset. Figure 4.14 shows the difference in offset between room temperature and hot and cold limits (125°C and −40°C) as a function of input voltage, for the conventional (noncascoded) op-amp.

4.7   AC OP-AMP NONLINEARITIES

Under alternating current (AC) conditions, mismatch can still be an issue. The most significant AC characteristics are bandwidth and gain frequency characteristics. If gain remains high as phase difference between input and output reduces at higher frequency, the result can be a high-frequency oscillation. Internal frequency compensation is built in to some commercial op-amps to increase the phase margin. This gives an intentional reduction in bandwidth, but maintains output stability.

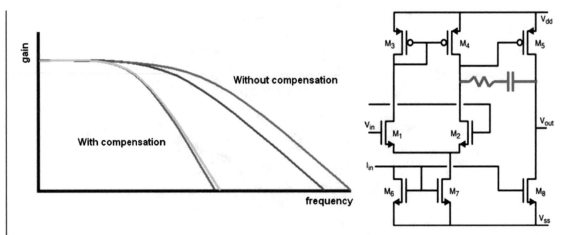

FIGURE 4.15: Circuit schematic showing typical placement of RC feedback. The RC not only reduces the maximum operating frequency but also, if properly designed, minimizes the variation in frequency cutoff. The graph shows with and without compensation frequency performance.

With increased frequency, gain reduces as a function of the intrinsic transistor characteristics and any additional parasitic component (capacitor and resistor) that make up the circuit layout. Because the circuit characteristics can easily be defined, it is typical to add compensation as part of the circuit design (Figure 4.15). This guarantees op-amp performance at the cost of reducing performance. The maximum frequency of operation is then defined by relatively stable RC combination, rather than the much more variable transistor and parasitic values.

The addition of an RC feedback path results in a lower bandwidth and a much more precise bandwidth. This is important as the phase difference between input and output can lead to oscil-

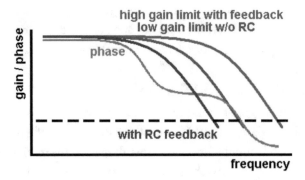

FIGURE 4.16: Graph showing with and without compensation frequency and phase performance. Note that only with the RC feedback does this configuration guarantee stability.

lation (Figure 4.16). If gain is still greater than 1 and the phase between input and output is 0, this results in positive feedback and oscillation.

4.7.1 Mismatch Issues in SRAM

Mismatch is of special concern in static random access memory (SRAM) cells. Each six- transistor (6-T) cell is self-contained and contains two PMOS and four NMOS devices. Being such a small cell, it is highly sensitive to mismatch. This is made worse since the cell itself is designed to be as small as possible. A small cell means that the transistors' widths and lengths are small, and mismatch is proportionally higher. There are three operational states for an SRAM cell: standby, read, and write (Figure 4.17):

Standby. If the word line is low, access transistors M_5 and M_6 disconnect the cell from the bit lines. The cross-coupled inverters formed by M_1–M_4 will continue to reinforce each other as long as they are disconnected from the outside world.

Read. Assume a 1 is stored at Q. Read cycle starts by precharging both bit lines to 1, then asserting word line WL, enabling both the access transistors. The values at Q and $\overline{Q}$ are transferred to the bit lines by leaving BL at its precharged value and discharging BL through M_1 and M_5 to a logical 0. On the BL side, the transistors M_4 and M_6 pull the bit line toward VDD, a logical 1.

Write. Apply value to be written to the bit lines. To write a 0, we apply a 0 to the bit lines, i.e., setting $\overline{BL}$ to 1 and BL to 0. WL is asserted and the value to be stored is latched on. (This works

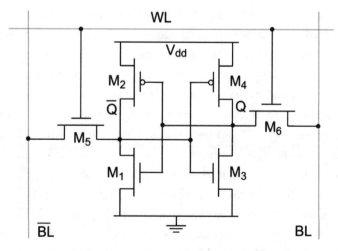

FIGURE 4.17: Conventional 6-T SRAM cell. The entire cell is made as small as possible, with the cell design generally centered around minimum size NMOS pull-down devices, long, but narrow PMOS, and pass-devices that are strong enough to overcome the pull-down of devices M1 and M3.

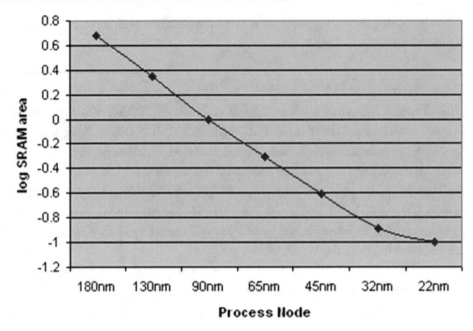

FIGURE 4.18: Cell area as a function of process node.

as the bit line input drivers are designed to be much stronger than the relatively weak transistors in the cell). Careful sizing of the SRAM transistors is needed to ensure proper operation.

Many SRAM arrays are up to or larger than 32 MB in size; hence, minimizing area is important. The smaller the SRAM cell size, the smaller the entire array. However, with such a large area of SRAM cells, we need to consider not only the local mismatch but also the sense amplifiers and read/write conditions must be set for variability variations. SRAM cell size and from that the individual component size is reducing approximately logarithmically with process node (Figure 4.18). As a result, over six process nodes of an area of an SRAM cell has reduced by a factor of 40. While the final data point on the curve does not follow the same curve as the previous six, there may still be further cell development at the 22 nm node that will improve the sizing of this cell.

Within a cell global variation is unimportant, providing the PMOS and NMOS devices vary at approximately the same rate; hence, if when NMOS devices are strong, PMOS devices are also strong (Figures 4.19a and b). The problem arises when PMOS and NMOS characteristics vary asynchronously, when the signal-to-noise margin is reduced (Figure 4.20) [3].

If we apply local mismatch to nominal models, we begin to see the effect of one-, three-, and six-sigma mismatch on the SRAM characteristics (Figure 4.19a). In normal circuitry, we only need to design using three-sigma, but SRAM is different. Regular circuitry averages several mismatches. In

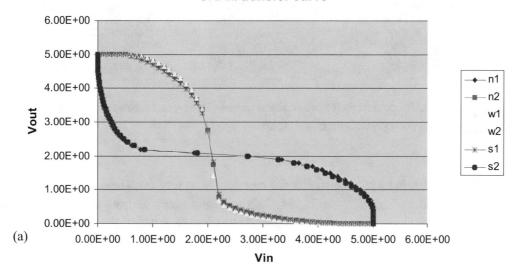

(a)

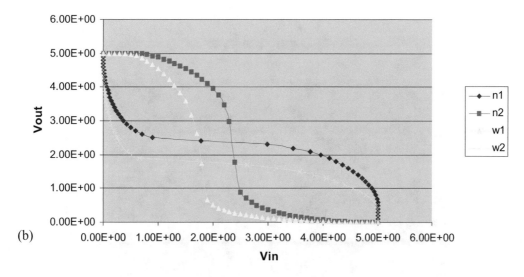

(b)

FIGURE 4.19: (a) Variation of the SRAM transfer curve showing little difference between SRAM cells when all models are weak, nominal, or strong at the same time (i.e., nominal NMOS, weak PMOS, or weak NMOS with nominal PMOS). (b) Variation of the SRAM transfer curve showing significant difference between SRAM cells when all models are not strength-matched.

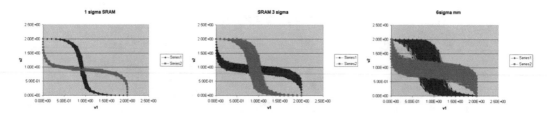

FIGURE 4.20: SRAM characteristics as a function of one-, three-, and six-sigma models, showing the reduction in the noise margin with increased sigma.

SRAM, basic cell has to be fully functional; the averaging is on very small devices. With an SRAM memory of 32 MB, we have (2^{25}, or 33,554,43p2) cells, and nearly all have to be functional (repair schemes that can be used for a small number of bits typically <100). Taking a look at the sigma corners:

Three-sigma=approximately 66,800 defects/MB (2 million/32 MB),
Four-sigma=approximately 6,210 defects/MB (200,000/32 MB),

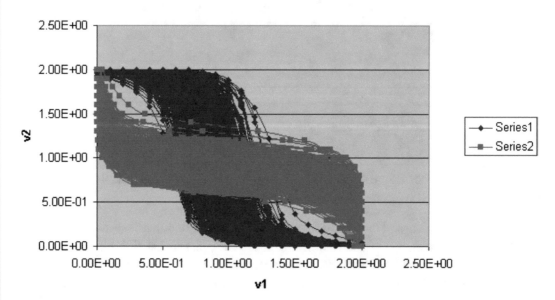

FIGURE 4.21: Combined effect of six-sigma mismatch and three-sigma global variability, showing that it is still possible to discern the SRAM state.

Five-sigma=approximately 230 defects/MB (7,500/32 MB), and
Six-sigma=approximately 3.5 defects/MB (110/32 MB).

Therefore, six-sigma cell performance is needed if we are to keep the performance adequate for the entire SRAM array.

4.7.2   Mismatch and Variability

A 64 MB SRAM has an area of about 8 mm×8 mm. As a result, across die variations affect variability and within the cell variations are local mismatch effects. What we have looked at above is mismatch on a nominal process, but as we saw earlier, corner models have an impact on performance. Figure 4.21 shows what happens when we combined the effects.

·   ·   ·   ·

CHAPTER 5

Lifetime-Induced Variability

5.1 END OF LIFE IN DIGITAL SYSTEMS

During the operational life of a microchip, transistors and interconnect are stressed in various ways. Some stresses do not have a significant impact until a catastrophic failure occurs, others can have a significant effect even in the early stages of degradation. The main transistor stresses that cause circuit degradation, ahead of catastrophic failure, are mismatch caused by negative bias temperature instability (NBTI) and stresses resulting from hot carrier injection (HCI) or channel hot carrier (CHC). Lifetime-induced stresses, where variation can occur in a circuit, are time-dependent forms of variability and/or mismatch.

These have to be handled in a similar way to mismatch, except that they are more difficult because, at time 0, there may be little or no mismatch; however, as time goes on, additional mismatch occurs.

5.1.1 Hot Carrier Injection

HCI is where an electron or hole gains enough energy to overcome the potential barrier between silicon and oxide, thus becoming a "hot carrier." Hot carriers can degrade the dielectric causing electron and hole trap formation; increase leakage modify the threshold voltage, prior to failing. HCI occurs when both gate and drain voltages are significantly higher than the source voltage. In digital circuits, this is a transient condition, but can be a DC condition in some analog circuits. The phenomenon results in high electric fields near the drain. These accelerate channel carriers into the drain's depletion region. HCI mostly affects NMOS devices. Most of the mismatch effects caused by HCI and NBTI are similar; however, HCI generally affects NMOS devices, while NBTI affects PMOS devices. As a result, we consider the effects in detail only for NBTI.

5.1.2 Negative Bias Temperature Instability

NBTI affects PMOS devices, particularly during negative gate stress (when the device is on, i.e., in linear operation). NBTI is the result of generation unsaturated silicon dangling bonds which form

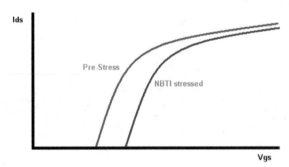

FIGURE 5.1: Representation of pre- and post-stress PMOS I_{ds}. V_t is affected significantly by NBTI, which also affects the I_{dsat} current.

interface traps. NBTI causes an increase in the threshold voltage and a decrease in drain current and transconductance (Figures 5.1 and 5.2a). There is some level of recovery the moment the stress is removed, depending on the length of stress; the longer the stress, the less the recovery.

V_t, on the other hand, has a more prominent and immediate effect, which is a concern for analog schemes. A 30-min stress has an approximately 5% degradation effect on V_{tsat}, and a 60-min stress increases the degradation by a further percentage point (Figure 5.2b). This indicates that NBTI stress has a more significant initial effect, which is a concern as this can lead to earlier failure than a linear or a late occurring catastrophic failure.

Although the effect of stress is significant (most analog circuitry relies on V_t matching to better than 1 or 2 mV), actual stresses in most analog circuits are significantly lower than the stress condition applied in NBTI stressing. Digital circuits are more significantly stressed, but they are less susceptible to performance reduction as their performance is a function of drain current, not V_t.

Effects on digital circuits of NBTI. What conditions are most significant for digital circuits resulting from NBTI stress? From Chapter 2, we have effective digital logic frequency:

$$\text{Frequency} = 2ptnt/c\left[nt(V_{dd} - V_{tlinp} + V_{tsatn}) + pt(V_{dd} - V_{tlinn} + V_{tsatp})\right].$$

However, only the PMOS is stressed by NBTI, hence:

$$\text{Frequency} \propto pt/\left[nt(V_{dd} - V_{tlinp} + V_{tsatn}) + pt(V_{dd} - V_{tlinn} + V_{tsatp})\right].$$

At high voltages, $V_{dd} \gg V_{tsat} - V_{tlin}$, thus:

$$\text{Frequency} \propto p/(nt + pt),$$

where $pt = I_{dtranp}$, $nt = I_{dtrann}$. Hence, at high voltages, frequency is independent of V_t. If $nt \approx pt$ (at time 0), the effect of 3% change in pt on frequency is about 1.5%.

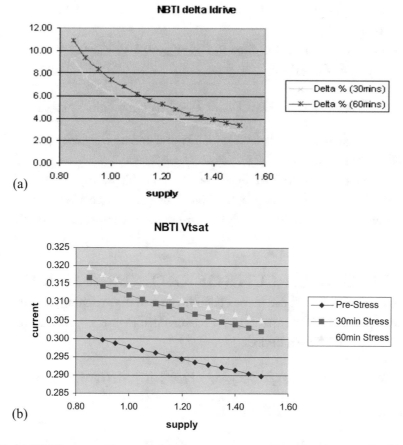

FIGURE 5.2: (a) NBTI stress effect on *I*-drive as a percentage plotted against supply. (b) NBTI stress effect on Vtsat as a function of supply voltage.

At low voltages:

$$\text{Frequency} \propto ptnt/[nt(V_{dd}-V_{tlinp}+V_{tsatn})+pt(V_{dd}-V_{tlinn}+V_{tsatp})].$$

If $nt \approx pt$ (at time 0), $V_{tlin}=V_{tsat}=0.3$ V, $V_{dd}=0.8$ V, and an increase in V_t in PMOS is 20 mV. If $nt=1$ mA, starting $pt=1$ mA and ending $pt=0.9$ mA. Hence, the starting frequency is α1E-3/[2(0.8)]=6.25E-4 and the ending frequency is α 9.0E-4/[1(0.78)+0.90(0.82)]=5.928854E-4, obtaining a frequency reduction of 5.138%.

Which is the largest effect: V_t shift or I_{tran} shift? Assume first an I_{tran} shift only. If $nt \approx pt$ (time 0), starting $pt=1$ mA and ending $pt=0.9$ mA, thus:

$$\text{Frequency} \propto ptnt/(nt+pt),$$

with starting frequency $\propto 1\mathrm{E}{-}3/2\mathrm{E}{-}3 = 0.5$ and ending frequency $\propto 9.0\mathrm{E}{-}4/1.9\mathrm{E}{-}3 = 0.474$, a frequency reduction of 5.2%.

Next, assume just V_t shift. If at time 0, $V_{\mathrm{tlin}}{=}V_{\mathrm{tsat}}{=}0.3$ V, $V_{\mathrm{dd}}{=}0.8$ V, and increase in V_t in PMOS is 20 mV at end of life, thus:

$$\text{Frequency} \propto 1/[(V_{\mathrm{dd}} - V_{\mathrm{tlinp}} + V_{\mathrm{tsatn}}) + (V_{\mathrm{dd}} - V_{\mathrm{tlinn}} + V_{\mathrm{tsatp}})],$$

obtaining starting frequency $\propto 1/(0.8 + 0.8) = 0.625$ and ending frequency $= \alpha 1/(0.82 + 0.78) = 0.625$, a frequency change of 0%.

Hence, substantially all NBTI-induced performance hit in logic circuitry is due to I_{dtran} changes, and V_t shifts may even marginally improve frequency. (This is due to threshold shifts that reduce the switching threshold for PMOS turn-on, partly compensating for the lower current switching).

5.2 EFFECTS ON CIRCUITS OF NBTI: CURRENT MIRRORS

We saw in Chapter 3 that V_t shift can make a big difference in the matching current of basic current mirrors. Referring to the PMOS current mirror schematic in Figure 5.3, NBTI stressing of current mirrors causes mismatch between devices.

$$I_{\mathrm{d}}(\mathrm{M}_2)/I_{\mathrm{d}}(\mathrm{M}_1) = [(W\mathrm{M}_2/L\mathrm{M}_2)(V_{\mathrm{gs}}\mathrm{M}_2 - V_{\mathrm{th}2})^2]/[(W\mathrm{M}_1/L\mathrm{M}_1)(V_{\mathrm{gs}}\mathrm{M}_1 - V_{\mathrm{th}1})^2].$$

If W and L are the same and V_{gs} are also the same:

$$I_{\mathrm{d}}(\mathrm{M}_2)/I_{\mathrm{d}}(\mathrm{M}_1) = (V_{\mathrm{gs}}\mathrm{M}_2 - V_{\mathrm{th}2})^2/(V_{\mathrm{gs}}\mathrm{M}_1 - V_{\mathrm{th}1})^2.$$

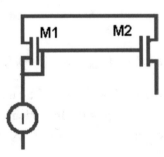

FIGURE 5.3: PMOS current mirror schematic. Generally, the drains will be at different voltages, so stress levels will vary.

A variation in V_t results in I_d changes, dependent upon V_{gs} voltage, assuming that $V_{gs} = 0.5$, $V_{th1} = 0.3$, and $V_{th2} = 0.30$ at time 0 and 0.32 after stress:

$$I_d(M_2)/I_d(M_1) = 1 \text{ (at time 0), and}$$

$$I_d(M_2)/I_d(M_1) = 0.18^2/0.2^2.$$

Hence, $I_d(\text{out})/I_d(\text{in}) = 81\%$ after stress.

The same techniques that give better current mirror matching can be used to achieve a state of reduced mismatch due to NBTI effects. A cascode current mirror, for example, has input and output drain voltages that tend to be within about 0.5 V of each other, thus NBTI stress will be similar.

5.3   EFFECTS ON CIRCUITS OF NBTI: OP-AMPS

A simple PMOS op-amp is shown in Figure 5.4. The devices most sensitive to NBTI in an op-amp are the input pair, where V_t shift can make a big difference in the matching. If the input pair is designated as transistor 1 and transistor 2, then because, at balance, the current in each transistor is equal, we have:

$$\text{Constant} = (V_{gs1} - V_{th1})^2 + (V_{gs2} - V_{th2})^2.$$

If all is balanced before stressing one PMOS, then the before vs. after:

$$(V_{gs1b} - V_{th1b})^2 + (V_{gs2b} - V_{th2b})^2 = (V_{gs1a} - V_{th1a})^2 + (V_{gs2a} - V_{th2a})^2,$$

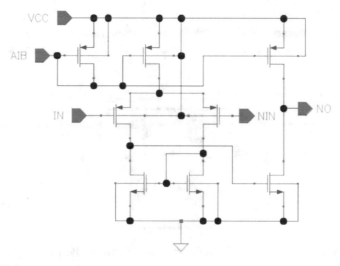

FIGURE 5.4: PMOS op-amp: the input pair is most susceptible to NBTI.

and if the before and after are the same for M_1:

$$(V_{gs2b} - V_{th2b})^2 = (V_{gs2a} - V_{th2a})^2,$$

thus:

$$V_{gs2b} - V_{th2b} = V_{gs2a} - V_{th2a}.$$

If V_{th2} after is 5 mV is higher than V_{th} before:

$$V_{gs2a} - V_{gs2b} = 5 \text{ mV}.$$

Hence, a linear relation exists for the effects of NBTI-stressed input devices. A 5 mV degradation of one PMOS input results in a 5 mV input offset.

Notes on offset. An offset of 1–2 mV between the input pair can be enough to prevent correct operation of a circuit. Consider a 10-bit flash ADC operating at 1 V:

$$10 \text{ bits} = 2^{10} \text{ quantization units} = 1,024,$$

$$1 \text{ V}/1,024 = 0.977 \text{ mV/step.}$$

As a result, an error of 2 mV in an input pair could cause 2–3 bits of error in some circuits and so must be carefully considered in designs.

5.4 EFFECTS ON CIRCUITS OF NBTI: STATIC RANDOM ACCESS MEMORY

The components in a conventional six-transistor (6-T) static random access memory (SRAM) cell that are most susceptible to stress are the PMOS pull-up devices (Figure 5.5). NBTI can affect

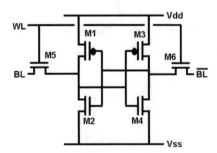

FIGURE 5.5: Conventional 6-T SRAM cell, showing the two PMOS pull-up devices: the NMOS drivers and NMOS pass gates.

these by causing an increase in the R_{dson} resistance. NBTI stress in an SRAM works in the following way:

Assume that the left-hand side of the SRAM (M_1 and M_2) has a high input and its output is low (the PMOS, M_1, is off) and that the right-hand side (M_3 and M_4) has a low input and its output is high (the PMOS, M_3, is on). Depending on the application, the SRAM cell could stay in this one state for prolonged periods—stressing M_3, but not M_1. Over time, M_3 becomes weaker, allowing the SRAM to gain a "memory" and preferentially switching to the opposite state if noise or switching events upset the state and increasing the "read failure probability" [4].

Is everything associated with NBTI degradation bad? No and yes. If the circuit still works, we do end up with a less leaky memory—which is usually considered to be good, but the variability is increased, which is not good for design. It is generally considered to be bad, but NBTI does not generally lead to degraded leakage.

·   ·   ·   ·

CHAPTER 6

Mismatch in Nonconventional Processes

Most integrated circuit designs use bulk silicon processing. Some other silicon-based processes are used for some applications, however, and it is worth briefly discussing these for the advantages and disadvantages they offer with respect to mismatch. We consider, in particular, two classes of process: partially depleted SOI (PDSOI) and fully depleted SOI (FDSOI).

Recent work from the Device Modeling Group at Glasgow compares the statistical variability in state-of-the-art bulk and SOI devices and studies various trapped/fixed charge densities in the devices. Their conclusion is that statistical variability in SOI is considerably lower than bulk [5,6].

6.1 WHAT IS SOI?

SOI [7] is a processing technique which, instead of creating components directly in a silicon substrate (see Figure 1.1), incorporates an oxide layer between the substrate and the area of 'active' silicon that contains the circuit components. This is generally considered beneficial as it reduces capacitance within the circuit and can have a number of other benefits that result in higher-performance and lower-power circuit designs. Figures 6.1 and 6.2 show the simplest form of SOI material and how components can be incorporated into the active layer, in what is known as a planar form.

If the active silicon region is relatively thick, the resultant components are said to be partially depleted. If the region is thin or if transistors are constructed vertically in the active region, the resultant components are said to be fully depleted.

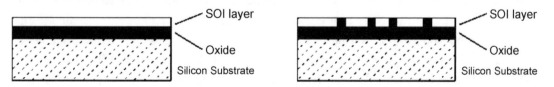

FIGURE 6.1: Left: A representation of an SOI wafer preprocessing. Right: Adding components (white areas), separated by oxide (black areas), gives islands or mesas of silicon.

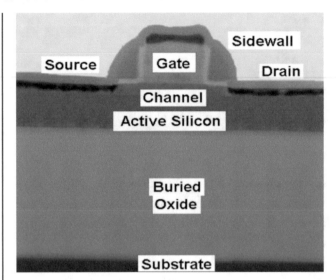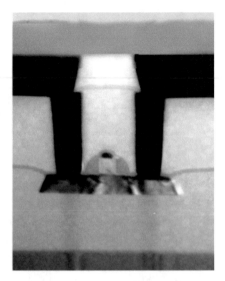

FIGURE 6.2: Image showing a close-up of a planar SOI transistor and a silicon island, surrounded by oxide.

6.1.1   Partially Depleted SOI

The term 'partially depleted' refers to the type of transistor created from a particular process, specifically, in this case, the fact that a MOS device built in thick active layered SOI has a body region that generally remains only partially depleted of carriers. Since this is not electrically connected through a resistive connection to the substrate or N-well, it is termed the floating body. As a result, there is an area of the body that can store charge. This is a potential problem, but it provides some benefits. The biggest problem as far as variability and mismatch are concerned is that depending on the charge in the body region at any given time, the device characteristics can vary. This can lead to phenomena known as the history effect, bipolar effect, and the kink effect, which are generally seriously detrimental to mismatch and variability.

Bipolar effect. Due to the parasitic bipolar device of the MOS device (in this case, the parasitic NPN of an NMOS) (Figure 6.3), where there is a floating, the capacitively coupled base is also leakage that affects base voltage. NPN turn-on can occur when the NMOS source and drain are high (e.g., in a transmission gate). This happens as the body drifts high until source, drain, and body are at same potential. If gate is low at this time, the circuit is 'off.' If the source is pulled low, then the base is pulled down due to base emitter (body source) diode turn-on. The parasitic NPN turns on until carriers are swept from the base region. This can take many picoseconds, during which current is drawn from the drain (collector).

Base charge can result in dynamic body voltage variation, which is a major issue in analog switching circuits.

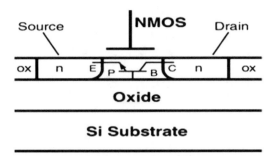

FIGURE 6.3: Bipolar effect in PDSOI; the parasitic NPN can turn on, discharging the floating body, but sourcing current from drain (collector) to source (emitter).

Kink effect. The kink effect refers to a discontinuity or 'kink' that can occur in the MOS *I–V* curve as voltage on the drain is increased. Figure 6.4 shows this effect (the actual effect is often not so pronounced and can just appear as a steeper slope rather than obvious discontinuity). The discontinuity is observed once impact ionization begins and is the result of an abrupt saturation current increase, which has the effect of raising the body voltage. In digital designs, it has the benefit of improving circuit performance; however, it is detrimental to most analog design.

History effect. History effect (Figure 6.5) is a gate delay change as function of switching history. Capacitive coupling and leakage modify the steady-state body voltage. At MOS, turned 'ON' body capacitively couples to drain, which is pulled low, possibly even below the source voltage level. Leakage makes it converge toward a DC value, but this can take several milliseconds. As a result, the subsequent switching can be based off a DC value or the voltage generated by an immediately preceding switching event that induced capacitive coupling. As a result, delay varies if next switching event occurs before body returns to equilibrium and as the input period is changed.

An issue with history effect, not directly connected to performance, is that it can perturb the expected leakage current. This leads to a temporary increase or decrease in leakage current. To

FIGURE 6.4: Kink effect in PDSOI: the parasitic NPN can turn on, discharging the floating body, but sourcing current from drain (collector) to source (emitter).

History Dependence:

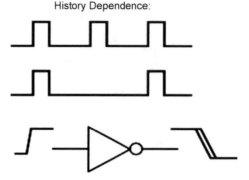

FIGURE 6.5: History effect is the variability of a switching signal, depending on the previous history of the switching signals through a gate or series of gates.

account for this, it is important to delay the measurement of leakage long enough to be sure that all the body voltages have settled. Usually, 1–10 msec is considered adequate to complete settling (Figure 6.6).

Calibrating body voltages. The simplest fix for the floating body phenomenon in PDSOI is to add a 'body tie' to components susceptible to mismatch (Figure 6.7). There are three ways to achieve this. The first is to tie the body to its own source. This can be done in most cases without requiring extra area (Figure 6.8). The second is to tie the body to the ground potential (for NMOS) or to supply (for PMOS). This requires additional silicon area in most cases. The third method to make a body tie is to the bodies of two or more devices. This method is particularly appropriate for input pairs of operational amplifiers, where matching is more important than absolute device performance. In this case, if the body voltage rises, the V_t will reduce on both body-tied devices. The

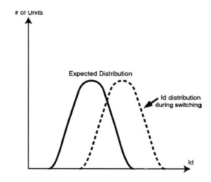

FIGURE 6.6: Diagram showing the expected shift in leakage distribution of PDSOI immediately following the termination of switching (dashed) and distribution after a long delay (solid).

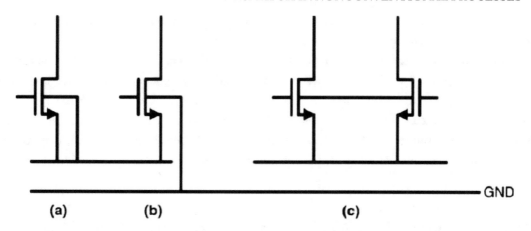

FIGURE 6.7: Schematic representation of body tie options. (a) Body tied to its own source; (b) body tied to ground (or in the case of a PMOS, to supply); and (c) body tied to another body.

body-tied-to-body example is particularly applicable to input pair thermal management schemes described below.

There are two associated methods of body voltage calibration that do not involve body ties: precharging and predischarging. In the former, just before a critical measurement is made, the body

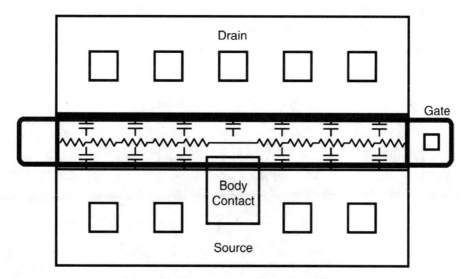

FIGURE 6.8: Method to create a body source tie, without additional area allocation for the device. The body contact is made through a silicided resistive contact to the contacts (square boxes in the diagram). The body contact is the same diffusion type as the body region, thus makes good ohmic contact to the body region.

or bodies are charged to the point where the diode that makes up the body to source voltage becomes forward-biased. The decay of the body voltage is similar enough between matched devices that for tens of microseconds, there is no significant difference in the body voltage and hence matching. Predischarging is the condition where, just before a critical measurement is made, the body or bodies are discharged to the point where the body and source voltages are equalized. Precharging and predischarging are achieved without an electrical contact to the body by manipulation of the gate, source, and drain voltages of the FET. While they can be efficient in terms of area, they require circuitry and design that will allow for possibly substantial calibration timing.

6.1.2 Fully Depleted SOI (FinFET)

There are few concerns of variability of FDSOI that are not also concerns of bulk silicon (already described) or are considerations of both FDSOI and PDSOI as discussed.

FDSOI comes in two forms:

- **Planar,** which is similar in concept to most PDSOI designs, but has a thinner active layer. Planar FDSOI may have additional variabilities due to difficulty in maintaining high accuracy on the active silicon layer thickness.
- **FinFET,** which creates transistor by etching through the active region to the oxide and leaving 'fins' of silicon, over which gate can be formed. If the fins are tall enough and close enough, it is possible to get a higher effective density of gate width than would be possible with a planar structure. Figure 6.9 shows diagrams and images of FinFET structures.

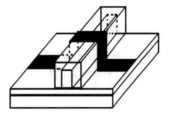
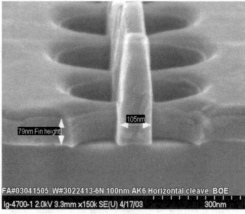
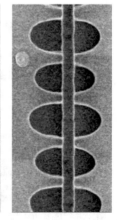

FIGURE 6.9: Views of FinFET structure, showing the troughs etched in the active silicon and how the gate wraps around the silicon fins to create a vertical structure.

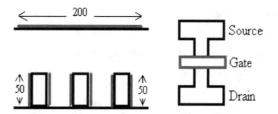

FIGURE 6.10: Representation of gate width achievable when using tall, closely spaced fins, compared to a planar FET.

FinFET active regions are calculated in a different manner to planar devices. The sidewalls become the active channel width/length. State-of-the-art fin width is 20–60 nm, while fin gate height of 50–100 nm is standard. It is possible to get a larger gate width per unit area than with a planar structure in some circumstances. Assume a fin height of 50 nm, a fin width of 20 nm, and pitch of 80 nm, then from Figure 6.10, it is evident that 300 nm of gate width can be squeezed into 200 nm silicon width.

6.1.3   Considerations That Affect Both FDSOI and PDSOI

Improved high-temperature operation is possible with SOI; leakage is two to three orders of magnitude lower than bulk at elevated temperatures. This is because leakage is due to lower forward voltage drop in diodes at high temperatures. As a result, typical bulk processes display reduced immunity to noise and an increased tendency to latch-up. SOI circuits do not have these parasitic

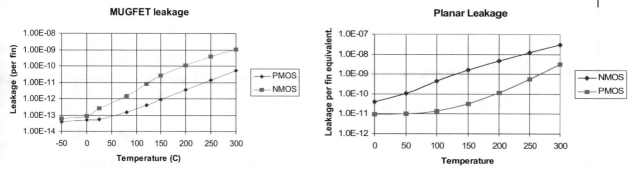

FIGURE 6.11: Leakage of FinFET (MugFET, left) technology and planar technology. Although leakage increases at the same rate in both technologies, because the fin technology leakage begins at a lower value, the devices remain viable to higher temperatures.

and thus avoid leakage to substrate. As a result, in SOI, the overall leakage is defined by the lowest leakage component in the circuit and not by some parasitic leakage path. For most applications, PDSOI can operate to about 250°C, very high temperature systems often benefit from the use of fully depleted silicon, when operation is possible up to 300°C or higher, while bulk operation is generally marginal even at 200°C.

NMOS and PMOS fin and planar leakage as a function of temperature is shown in Figure 6.11. Fin leakage increases by around 10× every 100°K temperature increase, with a leakage per fin of the NMOS device of 1 nA at 300°C. The PMOS has a leakage of 80 pA/fin at 300°C. This gives an I_{on}/I_{off} ratio of better than 6,000, which is a quite acceptable. Bulk planar leakage, meanwhile, also increases about 10× for every 100°K temperature rise. Its leakage equivalent for the NMOS is 30 nA at 300°C, and the PMOS is 4 nA at 300°C. The resultant I_{on}/I_{off} ratio for bulk is only about 220, which is marginal.

6.2   CIRCUIT EFFECTS OF HIGH-TEMPERATURE LEAKAGE

Application of the high-temperature operating option to circuits highlights some of the additional caution that must be taken due to temperature-induced variability. Leakage does not have a strong effect on digital circuitry until leakage is over 1% of the on-state current, but current mirror matching is a serious issue. Figure 6.12 shows current mirroring of an NMOS mirror. Matching is better than 3% over all conditions of bias and temperatures, except at very low currents, where mirroring is affected by leakage at high temperatures. This can be corrected by using a smaller current mirror, where a lower percentage of the current on the output device would be due to leakage.

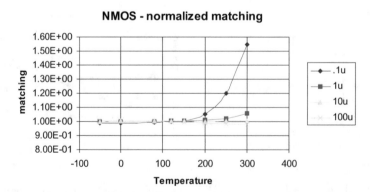

FIGURE 6.12: Mismatch of current mirror output as a function of temperature and matching current.

6.3   SELF-HEATING AND DISSIPATION PROBLEMS WITHIN A DEVICE

This is a form of variability, or mismatch, depending on the circuit configuration that has a faster response than lifetime stress degradation effects and slower than body effects of PDSOI. It affects mostly SOI devices, where removal of heat caused by current flow is not as efficient as in bulk silicon devices. The change in temperature in any given silicon island affects V_t and I_d characteristics. This is most prominent in analog circuits, which tend to have current flow consistently, rather than digital circuitry, where current flow is generally a transient phenomenon.

Figure 6.13 shows the difference in drive between a DC-driven (dashed) and a pulse-driven I–V (solid) curve. The DC-driven device heats up during the analysis, especially at high voltages and resistances, and shows a degraded current capability. While this is only generally a problem in large-power transistors in bulk material, it can affect all devices in SOI, which is why, in Spice simulations of SOI circuits, it is important to activate the thermal self-heating model within the device.

The input pair of an amplifier or a simple current mirror are good examples of where a matched pair may have different currents, and hence a thermally induced mismatch can occur. Schemes have been proposed that help to maintain the thermal matching within matched pairs [8]. In this patent performance, matching of devices in SOI are improved by thermally isolating matched devices within a continuous body of active material. Matched devices are isolated by an insulating wall of silicon dioxide (which surrounds the devices) and the oxide layer beneath, and they are arranged to minimize effects from external thermal sources (Figure 6.14).

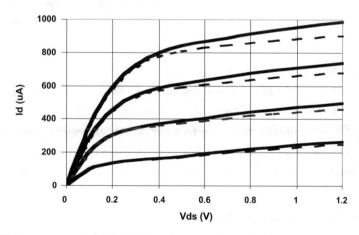

FIGURE 6.13: Drive currents with (dashed lines) and without (solid lines) self-heating. This shows an approximate 10% possible degradation of performance due to internal transistor heating.

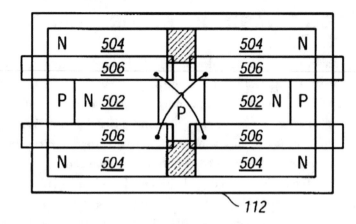

FIGURE 6.14: Proposed from Ref. [8] to match local temperature effects in SOI input pairs or current mirrors. It relies on combining all the devices in a single silicon island.

An example of the impact of thermal mismatch can be shown by considering the following example. When power dissipated in a current mirror differs between the input and output, for example:

A 100 μA mirror dissipates ~100 μA of V_t=0.2 V (20 μW) on the input.

A noncascoded mirror device may have a drain voltage of 1.5 V at 400 μA current, dissipating 600 μW at the output.

A local thermal resistance (junction to substrate) of 25,000°C/W results in a local temperature rise of about 15°C in the SOI island, compared to an insignificant rise in temperature in bulk material. This can lead to a V_t drop of 10–30 mV, which is a significant, transient local mismatch.

6.3.1   DC Heating From Elsewhere on the Same Chip

While the thick oxide around each silicon island makes it difficult for heat generated in one area of a chip to get to another, in some circumstances, an uneven heating can occur to two otherwise matched devices. The most likely way for this to occur is through thermal conduction from a 'hot' area of the chip to a 'cold' area of the chip through interconnect. It is normal practice to include exclusion zones above sensitive circuitry to mitigate the chances of thermal mismatch.

6.3.2   Transient or AC Thermal Coupling

Thermal time constants for adjacent devices on SOI are around a microsecond. Thermal transients in the output can affect the mirroring device dissipation and can couple to the reference device. Thermal gradients can introduce feedback with time constants that are long compared to electrical time constants.

A long thermal time constant introduces the possibility of unsimulated instability at frequencies less than 10–100 MHz, thus it is important to simulate thermal coupling in SOI circuits. This can be modeled in Spice simulation with the BSIM-PD and BSIM-FD models. Spacing devices further apart reduces intracell coupling; however, when matching for external thermal effects and process variation, increased separation is undesirable. One way to approach this is to apply the same scheme as suggested above for matching of local temperature effects.

6.4   DIGITAL CIRCUITS IN SOI

Historically, digital circuitry has rarely needed to consider self-heating effects. With SOI material, its higher current densities and reduced thermal conductivity of the buried oxide layer thermal effects are more pronounced, and may need to be considered, but are unlikely to be serious design issues.

Using the example of a ring oscillator as a simple digital circuit, as body voltage is increased, V_t decreases, leading to higher I_{drive} and leakage. As V_t varies, so does switching (typically 5–10% in PDSOI), as delay of a gate depends on recent history of terminal potentials (history effect). Analyzing Figure 6.15, however, we see that performance improvement at nominal supply (1.2 V, the

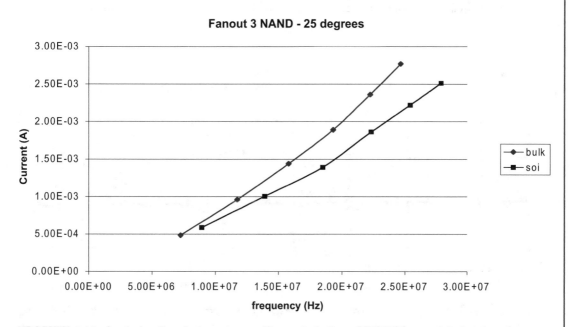

FIGURE 6.15: Analysis of equivalent ring oscillators in bulk and PDSOI material, showing that even accounting for an additional 5–10% switching variability due to floating bodies, we still have performance improvement.

third datapoint up each curve) is approximately 15%, meaning that we can design around the additional uncertainly in frequency and still achieve a performance improvement.

6.5   ANALOG CIRCUITS IN SOI

Within chip heating effects (from high to low thermal dissipation regions on the same chip) are well understood. These are likely to have to be considered as major mismatch concerns in SOI, but layout techniques have been described: simulation, including self-heating effects, can be done. These techniques also work with bulk, in addition to SOI.

6.5.1   Current Mirror: Kink Region Operation

A conventional current mirror in PDSOI has poorer response than bulk due to body voltage and thence V_t perturbation. Figure 6.16 shows a single-stage current mirror and associated waveforms for a PDSOI and tied gate (bulk equivalent) silicon mirror circuit. The SOI curve is steeper, representing poorer matching over supply voltage, but curves for both floating and bulk equivalent have poor mirroring.

Operation away from (in particular, below) the area of the kink discontinuity is especially important for analog circuits. We can achieve this simply in a current mirror, with the use of cascoding. A conceptual circuit to maintain the drain voltage at a low value is shown in Figure 6.17.

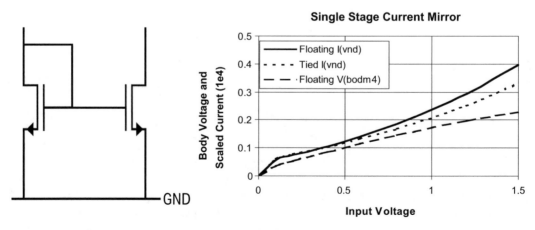

FIGURE 6.16: The single-stage current mirror (left) and associated waveforms for a PDSOI floating and tied gate mirror circuit.

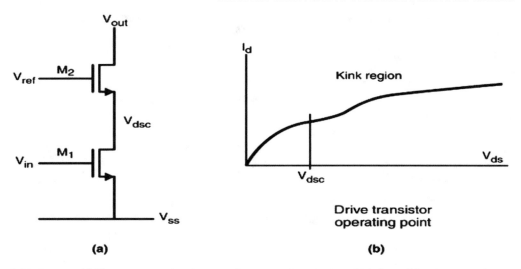

(a)

(b)

FIGURE 6.17: (a) The output side of a cascode circuit described in Chapter 2. This has the advantage of ensuing that the drain of the current mirroring device M_1 does not get high enough to be affected by any kink discontinuity. (b) Drive transistor operating point.

The result of this cascoding is shown in Figure 6.18, where it is seen that the mirror current of the floating and tied body cascoded mirrors is very similar. The floating body mirror remains very slightly worse than the tied body option, but it is likely to be adequately mirrored in most cases.

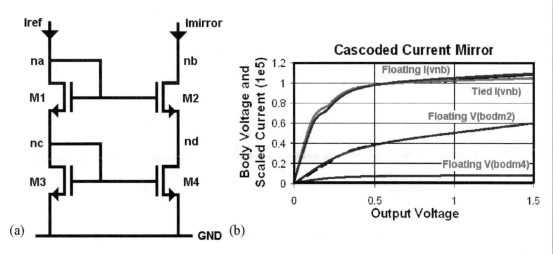

(a)

(b)

FIGURE 6.18: Current and voltage waveforms of PDSOI cascoded current mirror configurations, showing that circuit design can bypass the effects of the PDSOI floating body region.

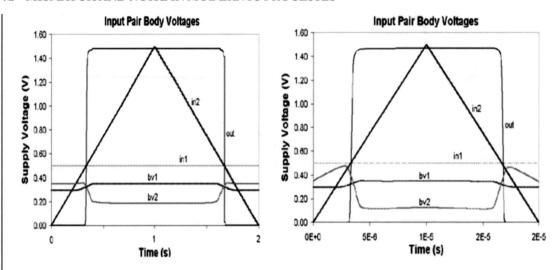

FIGURE 6.19: PDSOI CMOS operational amplifier graphs showing body voltages as a function of slow and fast input signals.

6.5.2   Operational Amplifier

The basic CMOS operational amplifier (Figure 4.2) requires a matched input pair for all but the least precision of uses. In PDSOI, supposedly matched devices may not be. As a result, the simple op-amp is of limited use with PDSOI technology. Figure 6.19 demonstrated the problem graphi-

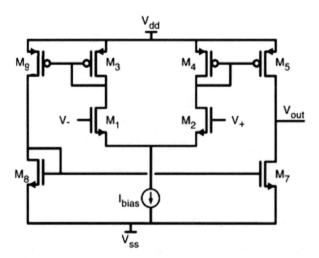

FIGURE 6.20: Operational transconductance amplifier, showing good input pair matching, but with mismatch on other the current mirrors.

cally. If the positive input voltage is ramped as it crosses a threshold of the negative input voltage, then the output switches from low to high. If all is balanced, all voltages are also balanced, and the body voltage of both sides of the input pair will also be the same. This is approximately the case for the instance where the input voltage ramp is very slow (*left-hand* side of Figure 6.19), but it is not the case when the input switches rapidly. The reason is that the body voltage of the positive input is being capacitively coupled by the rapid ramp. This results in an offset and hysteresis between rising and falling edges.

6.5.3   Operational Transconductance Amplifier

The basic operational transconductance amplifier design has reasonable component matching for balanced operation (Figure 6.20). This is particularly useful in SOI applications and for reducing mismatch in cases where channel modulation effects tend to dominate. In this amplifier type, the input pair voltages are matched, and under the circumstance where the V_{out} is equal to the V_t of M_8, the entire amplifier is matched. Generally, however, operation is required over a range of output voltages, and cascoding the mirrors M_3, M_9; M_4, M_5; and M_7, M_8 is desirable.

CHAPTER 7

Mismatch Correction Circuits

Mismatch can come in either circuit-induced mismatch, where matched circuit elements are not being operated in the same conditions, or random mismatch. Both can be designed around some area expense, but are generally dealt with in different ways.

Mismatch caused by circuit design has already been discussed for current mirrors and operational amplifiers in Chapters 3, 4, and 6. Correction for some of these mismatches has also been discussed. This chapter looks at methods to correct for random mismatch.

Random mismatch is generally detectable at final test either at probe or after packaging. Several methods have been developed to trim device characteristics to required values. Most of these developed initially as means to trim out global variation, as in the case of trimming reference voltages or frequencies into range.

7.1 TRIMMING METHODS

Trimming methods are generally digital, setting nonvolatile memory elements to a state that sets a digital or analog circuit to an appropriate level. Examples of circuits that can be trimmed include access codes, memory bit repair, unique identifiers, voltage trim, frequency trim, and leakage adjust. The newest applications include performance trim and leakage power optimization. Typical

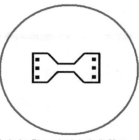

Exclusion Zone extends typically 50um

FIGURE 7.1: Plan view of a laser fuse. This can be of polysilicon or metal and is designed in a way that a short pulse of laser illumination will cut the fuse and change it from low to high resistance.

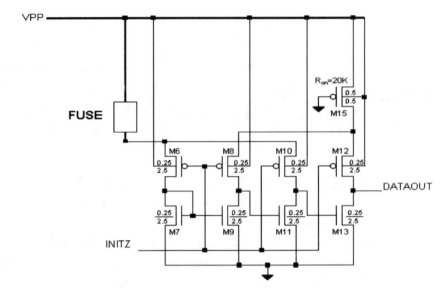

FIGURE 7.2: Fuse state detection is done using circuitry similar to this, where the state of the fuse is shown as a digital output during the time the init Z line is active.

nonvolatile memory elements used include laser fusible links, electrical fuses, EEPROM trim. and ferroelectric random access memory (feRAM) trim. These are explained below.

7.1.1   Laser Fuse

A fuse built from either polysilicon or metal can be used as a nonvolatile fuse element in a fuse farm or fuse array. An example of a fuse is shown in Figure 7.1. This type of fuse is generally restricted to less than a thousand due to the area requirements. All fusing is done at probe to trim frequencies and voltages. A fuse must be separated from any adjacent circuitry by several 10 sec of microns because when the laser evaporates the fuse, bits of metal or poly from the fusing process can end up far away from the fuse itself.

Electrically, the fuse must be measured. This is usually done at turn-on of a circuit, with the data latched into a D-latch or SRAM, from where the data is used during operation. The fuse sensing circuit is then turned off to save power. An example of the circuit used to sense the resistance of the fuse is shown in Figure 7.2.

7.1.2   E-Fuse Circuit

There are a number of advantages to designing an e-fuse instead of using a laser fusing method, but there are also some disadvantages to the technique. The e-fuse is beneficial if no laser is available on the tester, or if fusing has to be done after packaging is complete. It has disadvantages over the simpler laser fuse but in the circuit area needed to implement it and also in the reliability of the blown fuse (Figure 7.3).

The fusing system (left-hand side of the circuit) would, at first viewing, appear to be overly complex for a simple switch, so why so complex? It all goes back to variability of the fusing NMOS switch, which, as well as guaranteeing that fusing is done by the weakest possible NMOS, also has to not inadvertently partially turn on if the transistor is either at the strong corner, or if unpredicted logic switching occurs during turn-on. To prevent this, several protective features have been added:

1. The three logic inputs are designed to protect against digital input glitches by ensuring that inhibitions will be in place even if the main logic signal contains glitches.
2. The predrive stack, which separates the logic from the PMOS with a diode-connected MOS device, is to provide an additional diode drop of safety before the PMOS will turn

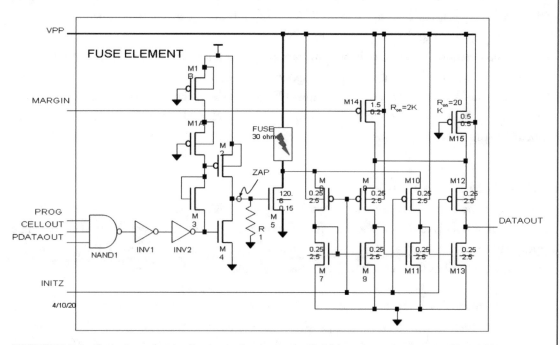

FIGURE 7.3: Complete circuit diagram of an e-fuse cell. Not only is the fuse itself needed, there is a large fusing and measurement circuit associated with the cell. The fusing portion of the circuit is on the left, and the measurement circuit is on the right of the fuse. The fuse is the block marked with a flash.

on. In case all other precautions for the logic fail, it gives a little more safety margin for the logic to stabilize and reach the expected value.

3. The resistor pull-down is added as an additional precaution—the PMOS not only has to turn on but also has to turn on with a current higher than that needed for the resistor.

These, while seemingly excessive, have provided the necessary protection to the fuse to ensure that the large transistor does not create a turn-on voltage glitch with potential to blow the fuse.

7.1.3 Sizing the Fusing Transistor

Assume a typical fuse resistance of 100 Ω and a fusing voltage of about 2 V. The fusing device might be expected to have to be about 10% of that (e.g., 10 Ω). For a typical transistor of current capability 1 mA/µm (I_{dsat}), 10 Ω in saturated conditions would be 50 µm wide. This looks good, but the problem is we actually need the 10 Ω in linear conditions (with 0.2 V across the device); thus, the 50 µm may be as much as 500 µm. This is way too big to be reasonable.

Fortunately, physics works in our favor. Because the fuse is much smaller than the transistor, the fuse heats up faster than the transistor. As the fuse heats up, its resistance increases approximately in proportion to the temperature in Kelvin.

The net result is that we only need to start in a condition where the fuse heats faster than the transistor. A safe condition for that is half supply; hence, we need to have our 10-Ω condition at 1 V, not 2 or 0.2 V, which means that a 100-µm-wide device should be adequate. In practice, because of variability in components and resistive drop within the chip, the transistor is actually sized to about 120 µm wide.

7.1.4 E-Fuse Sense Circuit

The right-hand portion of the circuit diagram in Figure 7.3 is the fuse state detection circuit. Theoretically, all it does is to detect whether the fuse is open or short circuit—something that should be measurable with a current source and an inverter, so why the complexity? Several things make this a more critical and difficult circuit to design than a simple voltage level detector:

1. Care must be taken that it can withstand the extra voltage applied during fusing.
2. Pull-down current must be applied, which will detect a fused or nonfused state, and be switchable, so that once the state is read, the fuse circuitry can be turned off to prevent operational leakage current and limit the possibility of regrowth of a fuse, which is more likely to happen if power is applied.

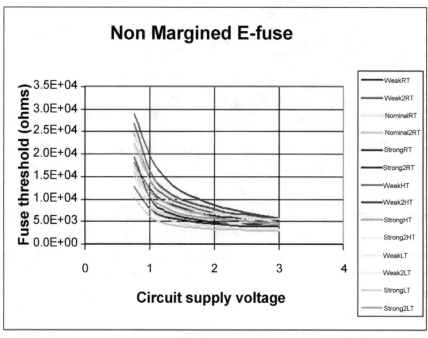

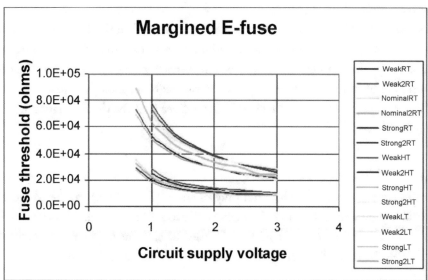

FIGURE 7.4: Comparison of fuses depending on whether the fusing verification was done with (lower) or without margin.

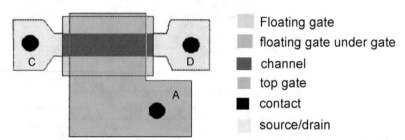

FIGURE 7.5: EEPROM cell, showing the floating gate that is generally lightly charged prior to programming and can start in either state.

3. Fused resistance is not generally infinity, as would be ideal, but around 20 kΩ minimum; and the maximum unfused resistance is not 0 Ω but close to 500 kΩ.

4. Because of these limitations, it is also necessary to verify fusing to 20 kΩ, but, in operation, check to a lower value (the margin) in case of regrowth. The operational limit is about 5 kΩ.

7.1.5   Importance of Margin

Why is it so important to include margin verification? If we fuse the fuse without verifying the margin and the test blown and unblown fuse values over an extreme of temperature and voltage conditions, although there is a tendency of fused structures to be higher resistance, there is no obvious differentiation of states (Figure 7.4, upper). If, on the other hand, we guarantee fusing to an additional margin and repeat the same experiment, we see that the differentiation of the states is clear (Figure 7.4, lower).

7.1.6   EPROM/EEPROM

An electrically erasable programmable read only memory (EEPROM) can be erased electrically, while an erasable programmable read only memory (EPROM) can be erased with ultraviolet (UV) light. Both are used for trimming. The EPROM is used when a one time programmation is needed, while the EEPROM is used when reprogrammation may be needed. Figure 7.5 shows an EEPROM cell design.

Unlike most other program cells, the big problem with EEPROM is that they do not start in a programmed state; hence, local variations can leave them in a '0' or a '1' state and even, in some cases, leave them in a state where the output can switch between a 1 and a 0. This can lead to some problems in schemes where a single time write is needed.

State fixing—the problem with EEPROM is that reprogrammation is possible. One way to prevent this is to use an electrical fuse to disable future programming; once the EEPROM is

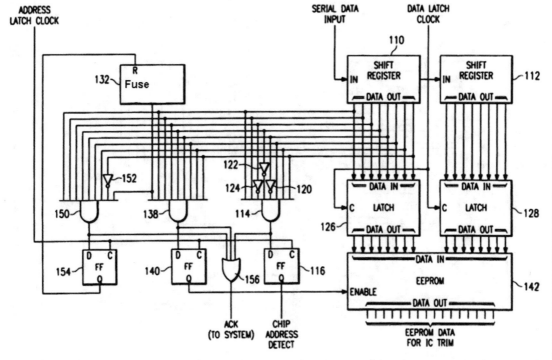

FIGURE 7.6: Block schematic for an e-fuse-controlled serial interface for an EEPROM array.

programmed, the fuse can be blown, which permanently disables programmation of the EEPROM cells (Figure 7.6). This permits the EEPROM to be programmed through a serial interface, until a specified data word is programmed into the serial data path that fuses the e-fuse and disables all future programming.

An alternative approach, suggested in 1995 [9] utilizes a probe pad, accessible only at probe test (Figure 7.7). This is used to set the EEPROM programming mode. EEPROMs are programmed, either at probe or final test (once packaged). Following programmation, a different program word is entered, which permanently disables the program-enable bit. The advantage of this scheme is that it does not require a large area e-fuse, and all the high current fusing leads associated with fusing an e-fuse. Its disadvantage is an extra probe pad is needed.

7.1.7   FeRAM Cell Design

The FeRAM cell (Figure 7.8) is based on a capacitor with ferroelectric (FE) dielectric, which becomes polarized with supply. This effect leads to charging differences which can be exploited in memory devices to create a nonvolatile memory (i.e., a memory device that does not lose its state even after being turned off).

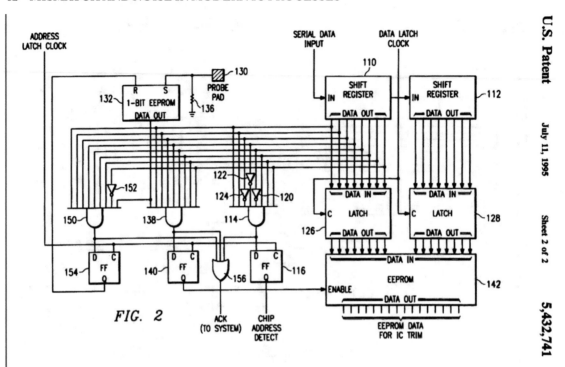

U.S. Patent   July 11, 1995   Sheet 2 of 2   5,432,741

FIGURE 7.7: Block schematic for an EEPROM-controlled serial interface for an EEPROM array.

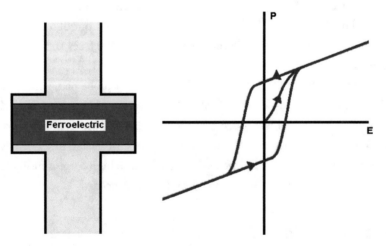

FIGURE 7.8: The Fe-capacitor used for the FeRAM looks like a capacitor, but with FE dielectric. The polarization (right) acts as a memory structure that can be combined with sense circuitry to act as a nonvolatile memory.

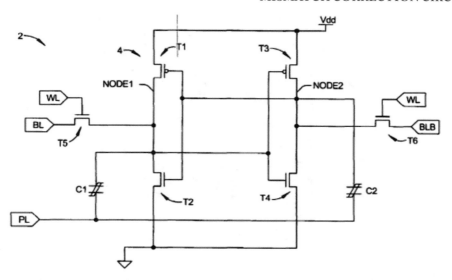

FIGURE 7.9: FeRAM latch circuit. This circuit is based around a conventional SRAM cell, but contains Fe capacitors that store a known state even when no power is applied to the circuit.

There are a number of ways that the FeRAM state can be detected, one that is useful for small quantities of FeRAM is the SRAM latch-based circuit (Figure 7.9). FeRAM—circuit design—latch is based on SRAM. Here, the 'plate' (PL) line is initially held at 0 V. The SRAM state is set as normal. Once the SRAM is set, the plate line is switched to supply and back. As a result, the memory cells flip to a positive or negative state during plate switching, giving the change in state needed by the SRAM for nonvolatility. After this, it is acceptable for power to be removed from the circuit. The Fe capacitors maintain their state even without power, and when the power is returned to the SRAM cells (generally at a controlled rate), the difference in state of the two Fe capacitors causes the SRAM to return to a known state.

7.2 CIRCUITS THAT CAN BE TRIMMED

A few examples of how digital trim can be used to modify analog circuits are shown below.

7.2.1 Frequency

The simplest way to modify the frequency of an oscillator is to adjust its current. This can be achieved by modifying the pull-up and pull-down currents of a stable multivibrator as shown in Figure 7.10. In place of a simple current source, say a current mirror, for the pull-up and pull-down currents, a binary weighted switchable current source controlled from the ROM source is added

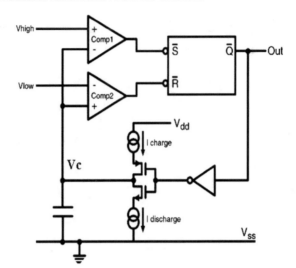

FIGURE 7.10: The simple oscillator can be trimmed most efficiently by adjusting the charging and discharging currents. This can be done by switching in and out current mirrors that act as charging and discharging currents.

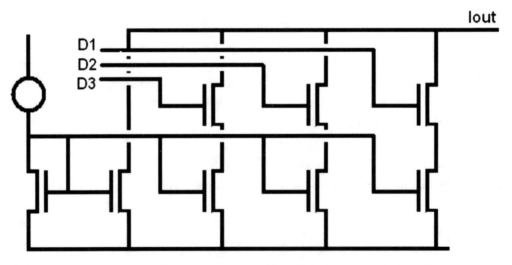

FIGURE 7.11: Digitally controlled current source. The digital inputs switch in the analog current sources. This configuration is usually binary weighted.

(Figure 7.11). Typical untrimmed frequency is accurate to about 10%. With trim, this can be reduced to less than 5%.

7.2.2 Voltage Trim

Typical voltage trim from a digital ROM input is shown in Figure 7.12. The digital trim is usually binary weighted, and trims in what might be a 5% variability to around 1–2%. The current source for the trim is usually derived from a bandgap circuit, and the resistance across which the reference voltage is developed is usually mirrored with the resistance in the bandgap to ensure matching of voltage over temperature.

7.2.3 Mismatch Trimming

Time 0 mismatch in a current mirror can be trimmed in the same way as shown in Figure 7.11, where transistors are turned on or off to get a closely matched input to output current.

Input pair current matching of mismatched pairs is more difficult because, normally, the extra circuitry loads down the amplifier to the detriment of its AC performance. A circuit that has been used for many years that unfortunately also causes significant loss in speed performance is the switching amplifier or chopper amplifier [10] (Figure 7.13). In the switching amplifier, time is taken out periodically to compensate the offset. The amplifier input is connected to ground and its polarity is switched between the normal and inverted states before detecting the resulting output DC

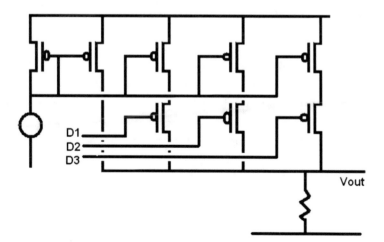

FIGURE 7.12: Typical voltage trim from a digital ROM input. The digital trim is usually binary weighted and trims in what might be a 5% variability to around 1–2%.

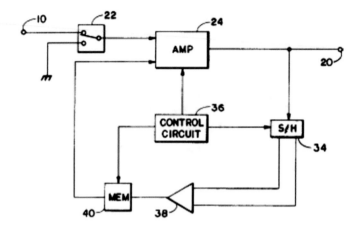

FIGURE 7.13: Switching amplifier of Lloyd Bristol, U.S. Patent 4,495,470.

levels. The difference between such output DC levels is used to generate an offset compensation signal to be applied to the amplifier.

A more recent technique that holds promise of being able to compensate for offset and not be detrimental to frequency performance makes use of body biasing. As seen in Figure 7.14, V_t can be adjusted significantly with body bias. In bulk material, this can be used on PMOS input devices or on

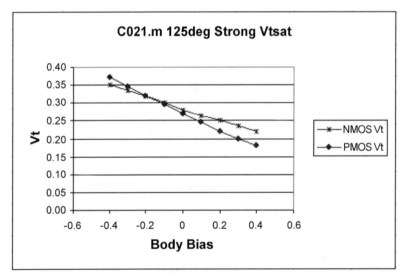

FIGURE 7.14: Variation of V_t with body bias suggests another means to adjust for time 0 mismatch.

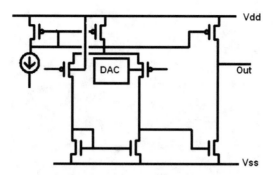

FIGURE 7.15: Body biasing of input pair to adjust for mismatch.

triple-well processes that isolate the NMOS devices. It is applicable to all PDSOI processes. Figure 7.15 shows how a digital trim can be used to modify the body bias and thus V_t of an op-amp input pair.

7.3   CASE STUDY: POWER IC DESIGN FOR TESTABILITY

This integrated circuit required a voltage accurate to 1% and a frequency accurate to 1.5%. With the technology available, local mismatches and variability did not permit this level of accuracy, so an EEPROM trim was used [11]. A serial interface was used to do the programming, as no additional pins were available on the package. This was achieved by setting an EEPROM-enable bit from a probe pad during wafer testing to 'enable' an array of EEPROM cells to be programmed at both

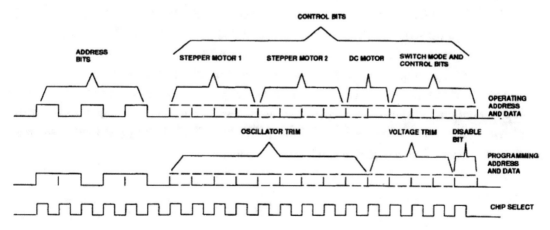

FIGURE 7.16: Example of serial input with separate address bits for operation and programming. A chip select/clock synchronizes the serial input data.

FIGURE 7.17: Die plot of the integrated circuit discussed in the power IC design for testability case study.

probe and final test through the serial interface using a different address to that used by the device for normal operation (Figure 7.16).

Following completion of the final test programming an EEPROM-disable signal is sent to the chip through the serial interface. This unique signal programs an EEPROM bit with a '0' and disables all future attempts at reprogramming the bank of EEPROMs. In the case of the oscillator frequency, the output from a counter-network connected to the oscillator is buffered to an open drain output and the frequency can be monitored. The chip layout for this is shown in Figure 7.17. The power outputs are shown in the lower half of the circuit, surrounded by guard-rings to help minimize substrate noise (see discussion in "Part 2: Noise" of this book). Most of the sensitive analog is in the upper left, and buffering and logic are in between.

PART II

Noise

CHAPTER 8

Component and Digital Circuit Noise

Noise is found in all integrated circuit processes and is present in all known or likely materials that may be employed in the future. For the next 10 or more years of expected silicon scaling, noise is going to be a major design concern [12].

Noise has to be separated into two forms: component and circuit. Component noise is generated within the component, while circuit noise is a result of coupling between circuits or components within a circuit.

8.1 COMPONENT/SILICON-INDUCED NOISE

There are several types of component noise as discussed below.

8.1.1 Thermal Noise

Thermal noise, also called Johnson noise or Nyquist noise, is generated by random thermal motion of electrons. This was first measured by John B. Johnson in 1928 and explained by Harry Nyquist. Thermal noise is applicable to any resistive conductive medium, which includes the channel of an on-state MOS device. It appears as a constant noise with respect to frequency and is the second most important noise source after $1/f$ in MOS devices.

8.1.2 $1/f$ Noise

$1/f$ Noise, also known as flicker noise, is highest at low frequencies and falls off as a function of $1/f$. In MOS devices, it is the result of populating and depopulating defect centers in the gate oxide. It is the most important noise source in MOS transistors.

8.1.3 Shot Noise

Shot noise is the result of random fluctuations of electric current in a conductor and is the result of charge being carried in discrete packets of electrons. Shot noise is an issue in bipolar devices, but not normally considered in MOS devices.

8.1.4 Burst Noise

Burst noise, also known as popcorn noise, consists of step transitions between levels, with a frequency lower than 100 Hz. Burst noise may be the result of random trapping and release of charge carriers at thin film interfaces, but numerous causes have been hypothesized.

8.2 PHYSICAL SOURCES OF NOISE

The channel of a MOS transistor is a significant source of noise. Traps in the oxide fill and empty randomly, affecting the V_t and current drive. Low-frequency, flicker (or $1/f$) noise is not just a tunneling-related phenomenon but also a thermally activated process. Figure 8.1 shows how the traps (circles in the oxide) may be distributed. Traps are filled from the channel, and emptying traps released into the channel region induces noise (Figure 8.2). The energy required for each of these traps to fill and empty may be different and can depend on the input conditions of the device.

An early empirical analysis of $1/f$ noise was undertaken in 1969 by Hooge [13]. This analysis has a tendency to underestimate noise quite significantly. An alternative early model came in 1957 from McWorter [14], whose model was based on surface charge trapping.

Most significant among device noise considerations is MOS transistor noise caused through impact ionization and oxide traps. Impact ionization effects are caused when the electrons from the source region (for an NMOS) are accelerated to the point where ionization can occur. Once a charge unit (electron or hole) has been trapped in the oxide, it can become a noise source, which will be more or less significant depending upon the trap depth within the oxide, its activation energy, and the temperature of operation of the circuit.

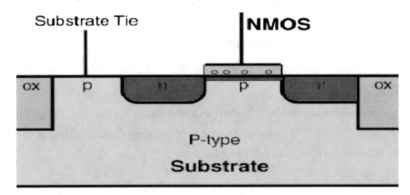

FIGURE 8.1: Demonstration of the random placement of noise centers in the oxide. The noise centers randomly fill and empty, modulating the channel. The variation in current at a given voltage is the measure of component noise.

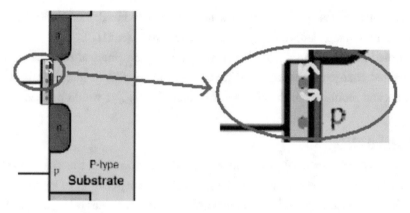

FIGURE 8.2: Charges from the channel are trapped and released by the oxide traps, leading to noise.

Only traps close to the Fermi level are likely to be active and thus contribute significantly to the noise level. As a result, noise varies with temperature. Tunneling of carriers within the oxide is most efficient when occurring between trap centers that are at the Fermi level energy.

While it is possible for tunneling to occur at other trap levels, any traps far away from the Fermi level are statistically likely to be filled or empty, dependent upon their energy position. Thus, tunneling far from the Fermi level is not significant enough to affect the overall noise characteristics.

Without successful prediction of noise, many circuits are at risk of being overdesigned. Static timing and noise analysis tools have demonstrated their importance in verifying digital designs.

8.3   NOISE SIMULATIONS

Spice simulators allow the analysis of noise. The MOS model, BSIM4, which is the most used Spice MOS model, provides several noise options as discussed below.

8.3.1   1/f Noise BSIM Modeling

BSIM4 has two 1/f noise models [15]. When the model selector f_{noiMod} is set to 0, a simple flicker noise model is selected which is convenient for hand calculation. If $f_{\text{noiMod}}=1$, a unified physical flicker noise model is used.

For the case of $f_{\text{noiMod}}=0$, the noise density is:

$$S_{id}(f) = \frac{KF\Delta I_{\text{ds}}^{\text{AF}}}{C_{\text{oxe}}L_{\text{eff}}^2 f^{\text{EF}}}$$

Note that this gives several hints on how noise may be reduced: increase of capacitance (C_{ox}), increase of effective channel length (L_{eff}), and reduction in current (I_{ds}).

The unified model, when f_{noiMod} is set to 1, is based on physical mechanisms of trapping/detrapping-related charge fluctuation in oxide traps, which results in fluctuations of both mobile carrier numbers and mobilities in the channel. In general, $f_{noiMod}=1$ would be used for simulations.

8.3.2   Thermal Noise

In BSIM4, there are two thermal noise models specifically to model the channel (there are also thermal noise models to cover other resistive paths, but these are generally of lower importance than the channel thermal noise). One of the channel noise models is a charge-based model, the other is the holistic model. These two models can be selected through the model selector t_{noiMod}:

$$t_{noiMod} = 0 \text{ (charge-based); and}$$

$$t_{noiMod} = 1 \text{ (holistic).}$$

For the charge-based model, noise current is given by:

$$\overline{i_d^2} = \frac{4k_b T \Delta f}{R_{ds}(V) + \dfrac{L_{eff}^2}{\mu_{eff}|Q_{inv}|}} \text{NTNOI}$$

where $R_{ds}(V)$ is the bias-dependent source/drain resistance, NTNOI is short-channel fitting factor, and Q_{inv} is a charge-modeling factor.

The 'holistic thermal noise model' additionally has variables that consider channel thermal noise through Gm and Gmbs as well as induced-gate noise with partial correlation to the channel thermal noise.

8.4   JITTER AND NOISE IN DIGITAL CIRCUITS: CIRCUIT EFFECTS

Component noise can affect circuits, but there are other noise effects also including:

- supply noise,
- ground noise, and
- capacitive coupling.

All these, and the device noise can cause circuit impacts that cause a process to perform below expectations. Considering the inverter in Figure 8.3:

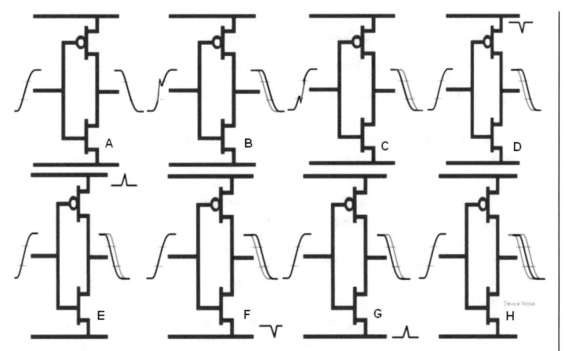

FIGURE 8.3: Effects of noise signals on switching response of digital inverter.

1. 'A' shows the expected input and output.
2. 'B' shows a positive-going glitch superimposed on a rising signal edge. This may be the result of capacitive coupling from an 'aggressor' signal line, and the result is a slightly faster signal transit time from input to output.
3. 'C' shows a negative-going glitch superimposed on a rising signal edge. This may be the result of capacitive coupling from an 'aggressor' signal line, and the result is a slightly slower signal transit time.

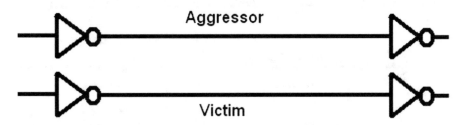

FIGURE 8.4: When two lines run close to each other, it is possible for one to induce noise onto the other during switching. The signal line causing the noise is the aggressor, while the line with noise being induced is the victim.

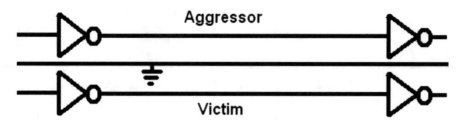

FIGURE 8.5: When a ground or supply runs between the victim and aggressor, the induction of noise onto the other during switching is greatly diminished.

4. 'D' shows a positive-going glitch on the supply line on a rising signal edge. The result is a slightly faster signal transit time. This can be due to an on-chip noise source or an external noise source.

5. 'E' shows a negative-going glitch on the supply line on a rising signal edge. The result is a slightly faster signal transit time. This can be due to an on-chip noise source or an external noise source.

6. 'F' shows a negative-going glitch on the ground line on a rising signal edge. The result is a slightly faster signal transit time. This can be due to an on-chip noise source or an external noise source.

7. 'G' shows a positive-going glitch on the supply line on a rising signal edge. The result is a slightly slower signal transit time. This can be due to an on-chip noise source or an external noise source.

8. 'H' shows the effect of noise on a rising signal edge. The result can be either a faster or slower signal transit time.

So how does a little bit of noise or jitter cause a process to be degraded? While for each of the noise-induced effects there is both the potential for a speed improvement as well as performance degradation, most designs have to be able to demonstrate functionality with either speed improvement or reduction, meaning extra design margin has to be included if circuit noise cannot

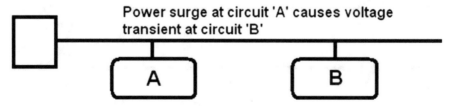

FIGURE 8.6: Voltage supply transients, caused by high-power switching of circuit 'A,' which affects all circuit supplies beyond the 'A' circuitry.

FIGURE 8.7: Voltage supply transient minimization by bringing a separate power supply for circuit 'B' to the bondpad.

be eliminated. The more noise there is, and the higher the voltage of the noise, the more degradation can occur.

Noise is induced from signal line to signal line by capacitive coupling between a low-impedance 'aggressor' line and generally high-impedance 'victim line' (Figure 8.4). This can be avoided by making sure the switching of the aggressor does not occur at a time that may affect the victim line or by placing a low-impedance, nonswitching line between them like in Figure 8.5.

Noise is induced through the supply line or ground generally as a result of power spikes along an internal supply rail. Figure 8.6 shows how this typically happens: Circuit 'A' is a high-power consumer, and following a power requirement surge, it can pull the entire supply line down, causing glitches on the supply at circuit 'B' further down the supply line. This is avoided by separating the supplies and bringing both back to the relatively lower-impedance bondpad (Figure 8.7). This is a technique used extensively in power-integrated circuits, where currents in the ampere range can be flowing around a chip at any given time.

In the case of ground noise, an additional mode for the noise to migrate across the chip is possible, although the method described for noise on the supply is also applicable for ground noise. Unless the process being used is an SOI process, it is possible to transmit ground-bounce noise through the substrate. If the substrate bounces significantly above or below the circuit, ground noise can be transmitted to any of several parts of the circuit (Figure 8.8).

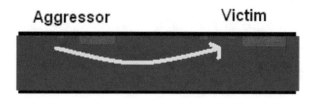

FIGURE 8.8: Noise coupling through substrate generally affects ground and body voltages. In extreme cases, it can cause latch-up in the victim circuit.

CHAPTER 9

Noise Effects in Digital Systems

In digital systems, noise manifests itself as phase noise. Another noise effect is jitter, the short-term variation of a significant instant of a digital signal from its ideal position in time. Both are effects of waveform-modifying interference on digital systems.

The reason noise manifests itself as phase noise is evident by looking at the effect of noise on a switching inverter in Chapter 8; noise has its greatest impact on digital performance when it affects the instant of switching. In 1994, John McNeill [16] presented a paper in "Jitter in Ring Oscillators." This covered a differential bipolar ring oscillator, but laid the framework for other work, notably "Jitter and Phase Noise in Ring Oscillators" by Ali Hajimiri, Sotirios Limotyrakis, and Thomas H. Lee [17] in 1999.

9.1 NOISE IN RING OSCILLATORS

Hajimiri et al. used a five-stage digital ring oscillator (Figure 9.1) to analyze the effect of phase noise as a function of currents and voltages. They noted that amplitude variation (outside the switching state) had no long-term effect on phase, but it was irreversible once a phase shift occurred (Figure 9.2).

Since most ring oscillators contain some kind of divider, it is difficult or impossible to resolve any amplitude noise on the signal, but phase change is seen at the output. If we assume that phase change is a single event, it can be measured as an absolute value in time difference between two

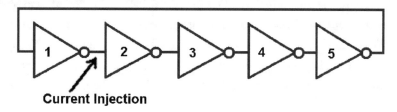

Current Injection

FIGURE 9.1: Ring of five oscillators, showing where current may be injected as a transient pulse, to add amplitude noise or cause a phase change.

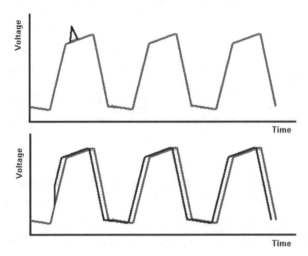

FIGURE 9.2: Upper trace: Noise injected into the nonswitching portion of the oscillator output just causes amplitude noise. Lower trace: the same noise injected into the switching portion of the waveform causes a phase change that does not correct itself.

cycles. However, generally, every rise and fall is affected in some way by a certain amount of phase shift, and so after passing through a divider, the actual phase shift seen will be the result of the sum-of-squares addition of all the phase shifts that have occurred during the cycle.

In the same way that noise can become averaged over multiple ring stages and divider stages, noise tends to be less of a problem in longer chains of gates, unless there is a correlation between the noise effects on the gate chain (caused, for example, by supply bounce).

9.2   DYNAMIC LOGIC

Attempts to minimize circuit-induced noise have been around for many years [18]. An early attempt based on a mixed signal chip discusses guard-ringing between digital and analog sections of a chip. The introduction to 'low noise digital logic technologies' states: "The achievable accuracy of many mixed mode ICs is currently limited by the adverse effects of digital switching noise. The noise arises from the large overlap current pulse that flows in the power supply lines when a conventional CMOS gate undergoes a state transition. In the layout of mixed signal ICs, it is common practice to use separate analog and digital power supply lines to minimize the common impedance, which, in turn, reduces the noise coupling between subsections." These statements remain valid to this day.

Historically, there have been several noise-tolerant techniques described in the literature, based on the use of a keeper for dynamic logic, precharging internal nodes and increasing source voltages. With ever-reducing geometries, affecting device leakage, and the capacitance per unit length of interconnect increasing, causing more noise coupling, there is a developing problem with dynamic logic circuitry [19]. The two design strategies suggested to address interconnect noise issues are:

1. To reduce the peak noise pulse generated in the interconnections through interconnect and circuit optimization (such as additional repeater insertion, wire and spacing sizing, and optimizing driver sizes); and
2. Designing noise-tolerant circuits.

   A conventional 'single-phase clock' dynamic AND gate building block is shown in Figure 9.3. This is clocked with the CLK signal, and inputs A and B. The fault-tolerant circuit of Mendoza-Hernandez [19] is shown in Figure 9.4. This works as follows (see Figure 9.5). On the falling edge of the clock signal, the circuit enters into the precharge state (stage 1). The dynamic node P_1 is precharged high, and the output is isolated from the inputs holding its previous value. After a time determined by the internal delay, the NCLK signal goes high (stage II), turning on Mn, which, in turn, precharges P_2 to V_{dd}. On the rising edge of the clock (stage III), the circuit is in transparent mode, allowing signal propagation. Stage 4 isolates the input from internal node.

   The major adder to noise immunity in this circuit is that noise on an input line cannot influence the output outside the transparency window. This is most useful when (1) noise is confined to

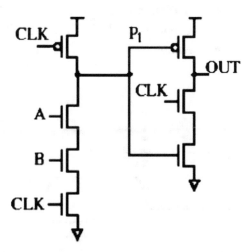

FIGURE 9.3: Conventional dynamic logic AND gate. A and B are the AND gate inputs, clocked by signal CLK.

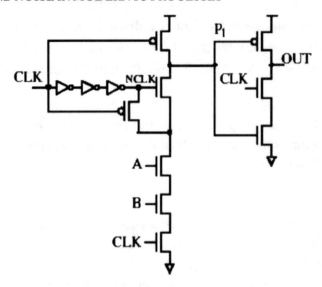

FIGURE 9.4: The fault-tolerant dynamic AND gate of Mendoza-Hernandez. This requires 15 components compared to 7 in the conventional dynamic logic circuit.

a section of the cycle outside the transparency window and (2) noise is restricted to signal lines and is not present on supply or ground lines.

Timing uncertainty is the most important noise-related concern in synthesized static CMOS logic blocks. Noise stability is also of some concern, because the restoring properties of static gates in the current technology generation tend to prevent propagation of noise. However, there is no restoring mechanism for delay uncertainties, and delay uncertainty can transform noncritical paths to timing critical. While designs using older technologies have largely been able to ignore noise in digital systems, below about 90 nm-technologies, noise in digital circuits has become an important design factor.

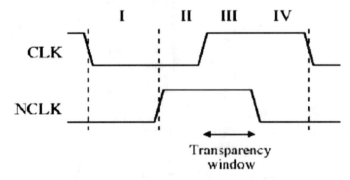

FIGURE 9.5: Operating, or transparency, window of Mendoza-Hernandez dynamic AND gate.

This has been considered [20] and resulted in the development of techniques to reduce noise in synthesized logic (logic that is generated by a CAD tool, not developed manually). Noise at the gate-level was considered during the net list generation stage using on-chip crosstalk models. The variation due to noise using a standard synthesis tool was shown to be about 18%. With synthesis incorporating crosstalk minimization, the uncertainty was reduced to below 3%, although an area and power penalty of 20% was observed.

The technique consists of driver-sizing limitations to prevent large fan-outs, which are more susceptible to noise. Placement and routing are done in a standard fashion, without timing or cross-talk optimization, then after layout RC extraction, static noise analysis is performed to monitor noise peaks, and noise-aware timing analysis is performed to monitor delay uncertainties. Routing data can be used for any additional post-layout optimizations. Traditional manual techniques are unrealistic for large digital circuits, and so this technique shows great promise for the future of deep submicron technologies.

9.3   INPUT PROTECTION

Signal inputs are usually protected, to some extent, from external noise with the use of input circuits that have hysteresis. Figure 9.6 shows a 'noisy' chip input signal and the output, if there is no hysteresis. A simple buffer, such as a double inverter (Figure 9.7), has little noise immunity, and so during switching, it is not unusual for a signal to switch through the threshold and causes multiple-switching events at the output. This can be alleviated, to some extent, with the use of hysteresis. Figure 9.8 shows the same input signal, when filtered through a hysteresis (or Schmidt trigger)

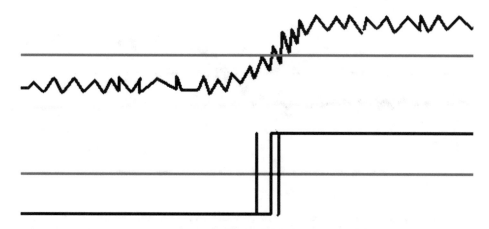

FIGURE 9.6: A 'noisy' chip input signal (*top trace*), and the output if there is no hysteresis. The output of the input buffer switches several times, which can disrupt the internal logic.

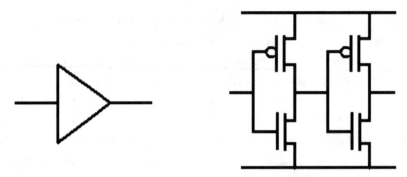

FIGURE 9.7: A no-hysteresis buffer symbol and an equivalent circuit diagram.

circuit (Figure 9.9). Here, at the expense of some switching delay, the 'on' threshold is delayed until the voltage is closer to supply, and the turn-off threshold is delayed until closer to the ground.

Another form of input noise that just about everyone must be familiar with is contact bounce. With contact bounce, a cheap metal contact switch or relay has a tendency to 'bounce' as it makes contact and repeat the requested operation multiple times, whether it is a calculator key, a TV remote, or the time advance on a digital watch. In general, switches take up to about 10 msec to close completely, during which time it may close and reopen several times (Figure 9.10).

A latch circuit, such as shown in Figure 9.11, is generally used to eliminate contact bounce, if a two-way switch is available. The latch works as follows: the switch starting state is 'up,' this means input A of N_1 is low ($A,N_1=L$), which automatically means the output is high (H) and therefore $A,N_2 = H$. Also:

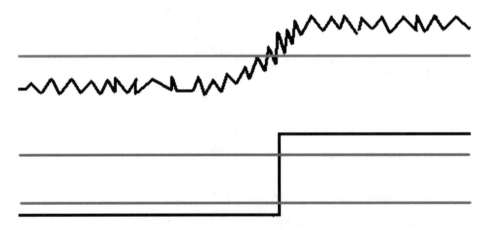

FIGURE 9.8: The same input signal, when filtered through a hysteresis circuit.

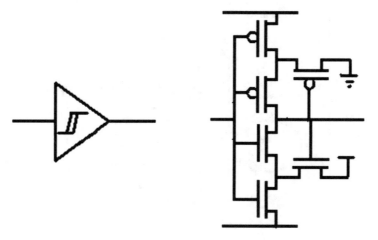

FIGURE 9.9: A with-hysteresis buffer symbol and an equivalent circuit diagram.

$$B,N_2 = H, \text{ so } B,N_1 = L.$$

When the switch switches to 'down,' the state changes as follows:

$$A,N_1 = H, \text{ since } B,N_1 = L, \text{ Output} = H, \text{ and } A,N_2 = H, \text{ thus output } N_2 = L \text{ and } B,N_1 = L$$

When the switch makes contact with its lower switch B,N_2, the node switches low; hence, the output of N_2 and B,N_1 switch are high. With both A and B inputs of N_1 now high and the output switches are low, A,N_2 then goes low.

With A,N_2 low, it does not matter if there is bounce on the low-switching node, as the output has switched low and will stay that way until the switch is returned to the original state. Both switching directions are debounced.

The two-way switch and flip-flop is an effective way to minimize switch bounce, but it has the drawbacks of requiring an expensive switch (relative to many of the simple metal-to-metal

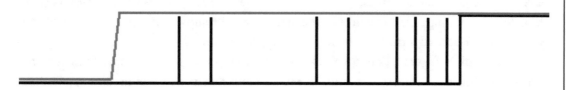

FIGURE 9.10: Contact bounce. The gray signal represents when the switch is pressed and the black represents the output conductance (open circuit to close).

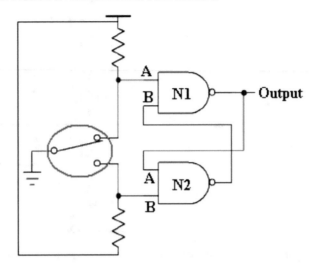

FIGURE 9.11: Contact bounce elimination using a two-way switch and flip-flop.

switches that are used in portable equipment). In addition, there is always current flow (and therefore power loss) in this type of design, another undesirable factor.

To resolve these problems, a monostable debouncer circuit is often used. Here, a monostable is activated on the first contact and is retained long enough to cover the period of switch bounce. Consider Figure 9.12: an input signal with bounce enters into a monostable. It also enters into a delay. The signal that emerges from the monostable is a clean pulse, designed to be longer than the switch bounce time. The signal that emerges from the delay circuit is the same as the input, but delayed long enough to ensure that the monostable output is the first signal to reach the OR gate. This is important because it takes a finite time to activate the monostable, and if no delay is incorporated into the initial signal, there is time for several spurious switching event to sneak through the OR gate before the monostable activates. Either signals into the OR gate being high will make the output of the OR gate high; thus, the monostable initially causes the OR output to go high. The input signal through the delay can turn on or off repeatedly before it settles into the on state and will not count as multiple-switching events. Eventually, the delayed input settles, meaning both OR gate inputs are high. At some time later, the monostable completes its cycle and turns off, but by this time, the delayed input has settled into its steady 'on' state.

The circuit in Figure 9.12 only provides positive-going debounce protection. The circuit can readily be modified to give negative-going debounce protection as well, and in most applications, this is necessary.

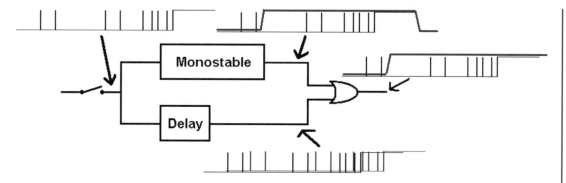

FIGURE 9.12: Monostable-type debounce circuit. This circuit is a one-sided debounce, as it only provides debounce for the low-to-high transition. The circuit can be modified to give bidirectional debounce protection.

This circuit has the advantage that only a simple on-off switch is needed. If a resistor is used as a pull-down and the switch is a momentary press and release type, the only time current is drawn by the circuit is when the switch is pressed, which is generally acceptable. For switches that need to toggle, it may be preferable to use a two-way type switch, which switches the input between ground and supply.

CHAPTER 10

Noise Effects in Analog Systems

Because analog circuits have much lower signal levels than digital circuits, they are inherently more susceptible to noise than their digital counterparts. Digital signal noise is generally easier to handle than analog noise, as it takes more noise to disrupt a digital circuit than an analog one, and some circuit topologies can help to desensitize certain digital circuits from noise events. However, digital circuits are more prone to generate noise due to their fast switching and rail-to-rail switching, which creates much larger voltage swings and higher capacitive coupling and rail swings. The combination of sensitive analog and high-current digital output circuits in the same chip is the worst design case.

10.1 SYSTEM ON A CHIP

This merging of technologies was investigated some years ago in a paper "Standard Cell Power IC Design" [21], where the use of analog standard cells was studied for reducing design time of mixed signal integrated circuits. A standard cell methodology was used to develop a complex mixed signal analog and digital power IC. The IC incorporated four-switch mode power supplies, five high-current full-H bridges, a computer-generated serial interface, and a 26-channel, 8-bit analog-to-digital converter. The chip is shown in Figure 10.1.

This early power IC process had 5 and 1O V CMOS logic and 5, 10, and 40 V analog/digital MOS, 40 V double-diffused MOS (DMOS), EEPROM, and bipolar NPN transistors. The paper concentrated on the advantages and disadvantages of a standard cell approach to the design process, but it also discussed the questions and solutions to combining an analog-to-digital converter (ADC) on the same IC as 10 A power outputs, without excessive switching noise and parasitic affecting the ADC precision. The path to a solution included an analysis of the timing of the switching of the power outputs, which operate at about 20 kHz, compared to a 5 psec conversion rate for the ADC. The ideal solution would have been not to do an ADC conversion during the output switching. The switching-mode (or switched-mode) power supplies (SMPSs), however, operate at around 200 kHz, which is a similar frequency to that necessary for the ADC, and therefore could not easily be designed not to switch during an ADC cycle, without introducing significant error into the SMPS output. Hence, additional guard-ringing was provided to protect the ADC cell and also the use of

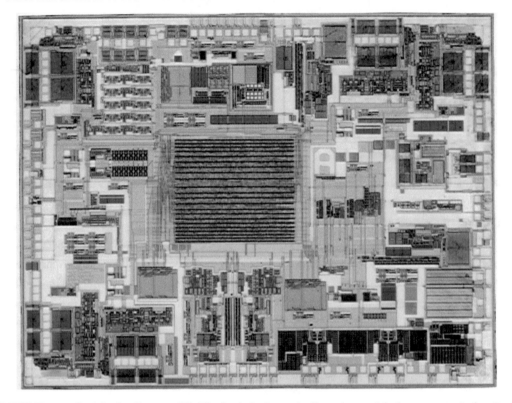

FIGURE 10.1: Standard cell power IC. The logic is shown in the center, with the power switches in the corners. Each output switch is surrounded by guard rings, but there was no requirement in this chip to isolate analog and regular logic.

Kelvin ground leads for the ADC to bond pads was implemented. These effectively minimized all the parasitic to an acceptable level.

10.2 NOISE IN AN OP-AMP

In an ideal noninverting op-amp (Figure 10.2), the equation that describes the output voltage is given by:

$$V_{out} = V_{in}(1 + R_f/R_1). \qquad (10.1)$$

If gain R_f/R_1 is high, this simplifies to:

$$V_{out} = V_{in}(R_f/R_1). \qquad (10.2)$$

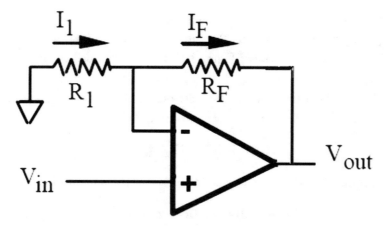

FIGURE 10.2: Ideal op-amp: noise analysis.

If V_{in} has noise associated with it, V_{in} with noise$=V_{in}+V_{n}$, then:

$$V_{outn} = (V_{in} + V_{n})\,(R_f/R_1).$$

Thus:

$$V_{outn} - V_{out} = V_{out}\,(\text{noise}) = V_{n}(R_f/R_1), \qquad (10.3)$$

showing that input noise is gained directly in proportion to the circuit gain. Note that this is under DC conditions and an ideal amplifier, at higher frequencies gain, will reduce and noise will reduce proportionally.

If the operational amplifier input pair happens to have the same level of noise (V_{ip}) as postulated for the input signal (V_{n}), then equation is similar:

$$V_{outnip} - V_{out} = V_{out}\,(\text{noiseip}) = V_{ip}(R_f/R_1). \qquad (10.4)$$

The feedback resistors can also be sources of noise although usually at a much lower level than active components.

If there is input noise and input pair noise, these combine as a sum-of-squares combination if they are uncorrelated; hence, noise would become:

$$V_{out}\,(\text{noisetotal}) = (R_f/R_1)\text{sqrt}(V_{ip}^{2}+ V_{n}^{2}). \qquad (10.5)$$

10.2.1 Effect of Noise Due to Two Cascaded Amplifiers

When two amplifiers are cascaded so that the output of one amplifier is fed straight into another to get extra gain, what happens to noise? Figure 10.3 shows a cascoded circuit with two amplifiers each of the same gain.

Assume that for the first stage, we have noise only in the input pair (i.e., the input signal is noise-free). This, from Equation 10.4 above, has an output noise:

$$V_{int} \text{ (noiseipa)} = V_{ipa}(R_{fa}/R_{1a}). \tag{10.6}$$

Again, from Equation 10.5 above, assume that the second-stage amplification is then:

$$V_{out} \text{ (noisetotal)} = (R_{fb}/R_{1b}) \text{ sqrt}(V_{ipb}^2 + V_{int}^2). \tag{10.7}$$

Combining Equations 10.6 and 10.7:

$$V_{out} \text{ (noisetotal)} = (R_{fb}/R_{1b}) \text{ sqrt} \{V_{ipb}^2 + [V_{ipa}(R_{fa}/R_{1a})]^2\}. \tag{10.8}$$

In order to minimize noise is it better to have gain in the first stage or second stage? Consider three cases:

1. first-stage gain=100, second-stage gain=1;
2. first-stage gain=10, second-stage gain=10; and
3. first-stage gain=1, second-stage gain=100.

All these give an equal DC gain of 100 and assume the same noise per stage, but what about the output noise?

Consider case 1 and apply to Equation 10.8:

$$V_{out} \text{ (noisetotal)} = \text{sqrt}[V_{ip}^2 + (100V_{ip})^2].$$

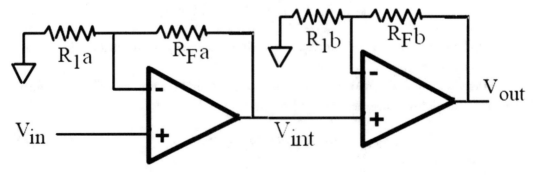

FIGURE 10.3: Cascaded amplification, demonstrating the effect of noise on a multistage amplifier.

Hence, V_{out} (noisetotal) = $100V_{ip}$.

Consider case 2 and apply to Equation 10.8:

$$V_{out} \text{ (noisetotal)} = 10 \text{ sqrt}[V_{ip}^2 + (10V_{ip})^2].$$

Thus, V_{out} (noisetotal) = $100.5V_{ip}$.

Consider case 3 and apply to Equation 10.8:

$$V_{out} \text{ (noisetotal)} = 100 \text{ sqrt}(V_{ip}^2 + V_{ip}^2).$$

Thus, V_{out} (noisetotal) = $141.4V_{ip}$.

It is clear from these cases that it is better to front-load the gain to minimize the overall noise. A rule of thumb for any amplifier chain is that the noise is almost completely defined by the first amplification stage, and so it is generally good practice to make the first stage with higher gain and lower noise at the expense of later stages in order to minimize noise.

10.3   LC OSCILLATOR

Figure 10.4 shows a typical LC oscillator circuit as used in many radio frequency (RF) and wireless designs. Phase noise is the biggest noise concern of LC resonant oscillators. The inductor and capacitor are generally integrated onto the IC, although the quality of the inductors can be quite low. At resonance of the LC, energy is added to the LC circuit every cycle, as needed, dependent upon

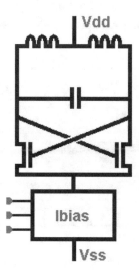

FIGURE 10.4: LC resonant oscillator circuit for RF or wireless applications. This type of oscillator is typically used in wireless or RF applications, where a stable reference frequency is needed of several gigahertz.

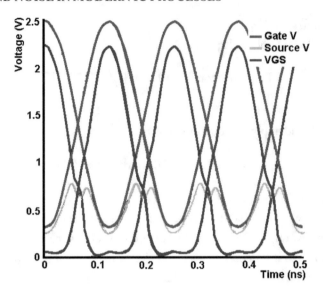

FIGURE 10.5: LC resonant oscillator waveforms, showing that the resonant circuit has top-up charge added at the least noise-impactive region of the resonance.

losses in the LC. Most of the energy in each cycle is resonant and transfers between L and C and back again. Figure 10.5 shows the gate and drain voltages. As can be seen, taking, say the left-hand device as the example, the transistor is 'on,' topping off the resonant charge, when the gate voltage is at its peak, which is the same time that the drain voltage is at its trough. This means the current charging of the resonator is at the sine-wave peak. This is a useful condition because it is the point

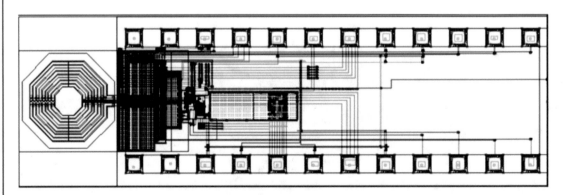

FIGURE 10.6: Digitally controlled LC oscillator layout: this is used as a performance verification module, the inductor is shown at the left, followed by capacitances of the LC oscillator, and cross-coupled transistor drivers.

on the cycle where any noise is least likely to impact the phase of the output. This is the same as in the digital oscillators described in Chapter 8, where noise during switching causes jitter. Figure 10.6 shows layout for this module as a test structure.

10.4 STATIC RANDOM ACCESS MEMORY

In Chapter 5, we examined the effect of variability and noise on SRAM cells. In many ways, noise has a similar effect to that of mismatch but does not generally show up as an issue at time 0, meaning it can cause issues at random during the life of the part, unless preemptive design takes into account the limits of noise.

As process technologies have advanced, giving even smaller components, the supply voltage has reduced alongside. In addition to the reducing supply trend, there are additional power minimization needs that mean SRAM cells are now routinely required to operate at even lower voltages, especially in standby states where the cell is not being written or read. Since low voltage operation reduces the static noise margin (SNM), additional care must be taken in the design and operation of modern SRAM circuits.

Ultra low voltage operation is crucial for modern SRAM designs especially for mobile and embedded applications. However, subthreshold current and threshold voltage mismatches developed in the memory cell are major concerns for low voltage operation.

The 6-T MOS static memory cell has two transistors and two load elements in a flip-flop configuration as shown in Figure 10.7. M_2 and M_4 are the driver transistors, M_5, M_6 are the pass gate NMOS devices, and M_1, M_3 are the load PMOS devices. Information is stored in the form of voltage levels in the flop-flop, which is formed by the two cross-coupled inverters. The flip-flop has two stable states typically designated '1' and '0.' The memory cell is embedded in an array of similar memory cells which are accessed by a row decoder which selects the word line (WL) and a column decoder circuit which selects the appropriate bit (BL) and bit-bar lines. Figure 10.8 shows a schematic for the 6-T SRAM with noise injection points for noise susceptibility analysis.

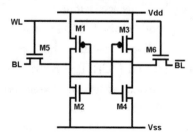

FIGURE 10.7: The 6-T SRAM cell cirucit diagram.

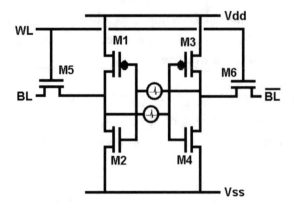

FIGURE 10.8: The 6-T SRAM, showing noise injection points for noise susceptibility analysis.

In logic gates, the noise amplitude and the noise duration are critical for dynamic stability. Dynamic noise margins are larger than the static ones because larger amplitude noise events can be tolerated if they persist for a sufficiently short time. Dynamic stability analysis can be applied to an SRAM cell resulting in an analytical model for evaluating its dynamic noise margin while in standby. This assumes that the noise source is a current noise pulse injected into the node storing a '0.' The dynamic analysis assumes that for a given noise amplitude, the minimum pulse duration for the noise to flip the cell's state can be determined. This is termed the 'critical pulse.'

10.4.1   Radiation Hardness in Static Random Access Memory

'Radiation hardness' is an expression of the resistance of a circuit to radiation-induced errors. Soft error rate (SER) is the measure of radiation hardness and is a function of many circuit, material, and environmental factors, including the type of radiation, supply voltage, and layout details. Memory cells are particularly susceptible to radiation-induced effects because, unlike many other circuits, a radiation-induced error results in a changed digital state on the memory cell. SRAM are not immune to radiation-induced effects, and this is normally considered to be a form of a system noise.

The net charge imbalance QCRIT necessary to upset the cell state is equal to $C\Delta V$, where C is the capacitance seen at the storage node of the SRAM cell and ΔV is the cell differential voltage (supply voltage in a 6-T SRAM Cell). For a 0.5–μm technology, the QCRIT is around 25–30 fC [7]. Cell size shrinks with technology, and QCRIT also reduces. Due to the presence of buried oxide, it is very difficult for the alpha particles to get injected into the channel. The reduced cell size and storage node capacitance improves cell performance due to a reduced number of parasitic. The sublinear improvement in cell access time is traded against the almost exponential deterioration in

SER. Alpha-induced SER is expected to surpass the cosmic ray (which is relatively insensitive to QCRIT, and SER remains relatively constant as technology scales) and become the major failure mechanisms for technologies 0.15 μm and below [7]. In conventional bulk CMOS, α-generated charges are collected mainly by the funneling effect, when particles collide with the drain diffusion layer. For bulk SRAM, the scaling of the cell and supply voltage reduces QCRIT, and the SER rises almost exponentially.

. . . .

CHAPTER 11

Circuit Design to Minimize Noise Effects

In many cases, noise within a circuit or on a chip can be minimized with appropriate design. In some cases, noise can be averaged out, and chip level noise, caused by coupling through the silicon or interconnect, can be shielded. When the noise occurs through the substrate, a technique called guard ringing is often used.

11.1   GUARD RINGING

The methods used to reduce this type of noise are to ensure very close connection of ground and body connections at a local level. This is usually effective in minimizing latch-up. Another method that is effective at the cost of some area is the incorporation of guard rings either around the victim circuitry or around the aggressor circuit. The guard ring is generally connected to higher voltages than the victim circuit, generating a large depletion region and channeling current flow preferentially to the guard ring (Figure 11.1).

All the prior noise sources are to some extent controllable, and designs can be modified as the noise sources are either a result of the devices themselves or due to noise coupling from one part

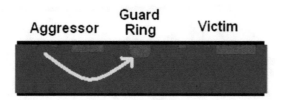

FIGURE 11.1: Guard ring to minimize the effects of substrate noise. Other techniques used in reducing the effect of substrate noise are improved victim body/ground contacts.

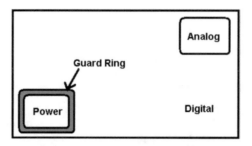

FIGURE 11.2: Representation of a guard ring view for reducing noise transmission across a chip.

of the circuit to another. There is, however, an additional source of noise that comes from outside a circuit, which is noise coming in on any connection into the chip. Inputs are usually protected to some extent with the use of input circuits with hysteresis. Supply noise is often minimized with a capacitor outside the chip.

A plan view of a guard ring surrounding a power aggressor component is shown in Figure 11.2.

11.2   NOISE SUPPRESSION THROUGH INTERCONNECT

As discussed in Chapter 8, noise is induced from one signal line to another by capacitive coupling between a low-impedance 'aggressor' line and a generally high-impedance 'victim line.' This can be avoided by making sure the switching of the aggressor does not occur at a time that may affect the victim line or by placing a low-impedance, nonswitching, line between them. Electrically, this looks like the combination view/schematic in Figures 11.3 and 11.4. The addition of a low-impedance line works because the C–R time contant reduces the amplitude of the signal substantially more than the victim line; the additional separation also reduces capacitance.

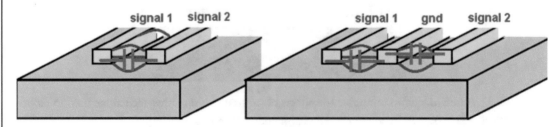

FIGURE 11.3: Diagrammatic representation of capacitive coupling between interconnect lines, without (left) and with (right) a ground separator (see also Figures 8.4 and 8.5 for the layout plan views).

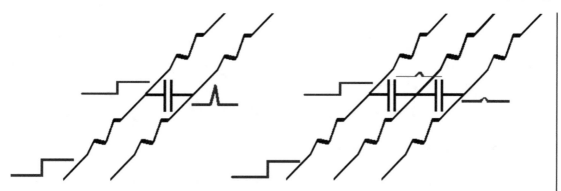

FIGURE 11.4: Schematic representation of capacitive coupling between interconnect lines, without (left) and with (right) a ground separator (see also Figures 8.4 and 8.5 for the layout plan views).

11.2.1   Supply Line Noise

Noise introduced through the supply or ground lines generally was discussed in Chapter 8, where it was shown that the resistance of the interconnect causes a voltage drop, which can affect circuitry. A technique that is frequently used was also explained in the same chapter, which is using separate power or ground lines to a common supply pin. This limits the voltage reduction to one line and the one circuit that is pulling the excessive current. Another technique that is used is to have specially designated interconnect layers that devoted to the high-current interconnect. These are usually placed on the top one or two levels and are of thicker metal.

11.2.2   Substrate Noise Reduction

If noise is introduced into interconnect through the substrate, it can be handled in two ways. The first and simplest is to implement a vertical version of the aggressor to victim coupling of Figure

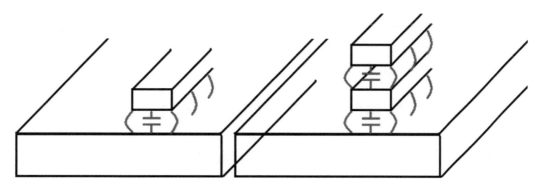

FIGURE 11.5: Shielding of signal line from potential ground bounce.

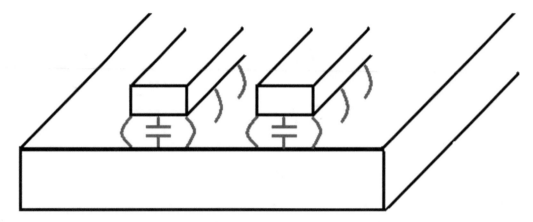

FIGURE 11.6: Use of differential signals to null out noise.

11.4. This is shown in Figure 11.5 and involves shielding the signal (victim) line, which is on metal 2, with a fixed, low-impedance line on metal 1, reducing capacitance in a similar manner to the horizontal shielding of the previously discussed noise reduction.

Another method that can reduce the effects of substrate noise, without requiring shielding, and with some other advantageous properties is the use of a differential signal. Here, two signals are sent across the chip, substrate bounce couples equally to the two signals, and when the signal is extracted at the receiving end of the circuit, the spurious noise signal on both signals is nulled out as the two signals are subtracted. This is particularly useful when designing analog systems (Figure 11.6). The block circuit diagram is shown in Figure 11.7.

This circuit is readily implemented. Assume first that a digital circuit, a single-ended input, can be converted into a differential input using an inverter. Conversion back to a single-ended signal can be either digital or can use a differential amplifier. For an analog signal, a single-ended input can be converted into a differential signal with the use of a unity gain-inverting amplifier, plus the input

FIGURE 11.7: Circuit diagram showing use of differential signals to null out noise.

signal to the unity gain-inverting amplifier. The conversion back to a single-ended output is done with an op-amp.

11.3   REDUCING COMPONENT NOISE

While there are several process-related means to control noise, the control most readily available to a circuit designer are electrical design characteristics, this includes the operating conditions (currents and voltages) and MOS device area. Indeed, most of the derivations for $1/f$ noise generally show a direct dependence on the square of current and an inverse dependence on MOS width and length (and an inverse dependence on capacitance). Device area noise dependency is easily understood by considering a single oxide defect-trapping center within a MOS device (Figure 11.8). As the device gets larger, the effect of the defect center is reduced.

Most noise models include this dependency, allowing the development of optimized noise reduction. In addition to noise effects, most models also allow statistical variability and mismatch to be included in order to obtain a comprehensive simulation of the device sizing with all these statistical effects incorporated.

11.4   CIRCUIT EFFECTS AND NOISE

11.4.1   Smart Sensors

Smart sensors are systems in which sensors and dedicated interface electronics are integrated on the same chip or in the same package. Examples include torsion/stress measurement, temperature sensing, wind velocity, and magnetic field analysis. Due to the low-level analog output of typical sensors, designing interface electronics that does not impair sensor performance is a challenge, where inherent precision is limited by $1/f$ noise and component mismatch. However, since most sensors

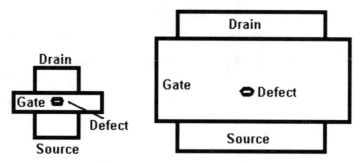

FIGURE 11.8: Demonstration of how a single defect has a larger effect in a small device than in a larger device.

are slow compared to the surrounding circuitry, dynamic techniques, such as are auto zeroing, chopping, dynamic element matching, switched capacitor filtering, and sigma–delta modulation can often be used to achieve higher precision.

11.4.2   Operational Amplifiers

Operational amplifiers are common in analog circuit design, and the methods described in the literature to minimize or reduce noise are numerous. One method to reduce noise is the use of a chopper amplifier [22]. Chopper amplifiers are DC amplifiers; however, some small DC (or low-frequency) signals that need high amplification can be difficult for a DC amplifier to work with due to inherent DC offsets and noise. It is much easier to build an AC amplifier where the DC offsets can be nulled. A chopper circuit is used to break up the input signal so that it can be processed as if it were an AC signal then integrated back to a DC signal at the output. In this way, extremely small DC signals can be amplified [23].

  A chopper amplifier (sometimes called a nulling or auto zero amplifier) works as an amplifier for awhile then corrects its offset voltage and goes back to working as an amplifier again, typically between 100 Hz and 20 kHz [24]. A chopper amplifier transfers its low-frequency noise and DC offset to higher frequencies by the use of modulation. Modulation is often implemented with MOSFET switches that are opened and closed alternately. Thermal noise and low-frequency $1/f$ noise are the main limits on low-noise amplifier design, and the shifting of noise from a lower to

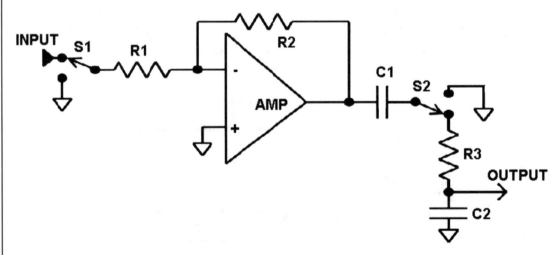

FIGURE 11.9: Typical simplified chopper amplifier design. The design achieves the dual advantage of eliminating mismatch-induced offset and reducing low-frequency noise.

higher frequency can improve the operating band performance. The use of chopper amplifiers can both effectively reduce the low-frequency noise of an amplifier and improve local mismatch, but with the tradeoff of higher operating current and lower bandwidth. Figure 11.9 shows an example of a simple chopper amplifier [24].

Its mode of operation is as follows: S_1 begins in the up position, allowing the input signal to be amplified to the output of the amplifier. C_1 does not pass a DC signal, which has the effect of nulling out DC offsets of the amplifier. C_1 and R_3 act as an RC filter, and as C_2 charges, the output rises to the point of the amplified input signal. However, this cannot remain the active state as the capacitor C_1 will discharge and the output will eventually leak to zero, thus periodically S_1 and S_2 switch simultaneously. In the state where S_1 is switched to ground, any amplification caused by input offset will be on the left-hand C_1 plate. The right-hand plate is then shorted to ground, so the offset value is stored on the capacitor plate. When S_1 and S_2 return to their original state, the gain at the output of the amplifier is the gained input plus the gained mismatch offset, but the output at S_2 is the gained signal only, without the offset component. The procedure is repeated at a frequency that is defined by the input characteristics, power consumption requirements, and component sizing. The switches can be constructed as transmission gates or other appropriate elements, but care must be taken that the gates used do not themselves introduce additional noise or offset.

.   .   .   .

CHAPTER 12

Noise Considerations in SOI

The main differences in the way noise is dealt with between bulk and SOI material is related to the SOI layer and component to component noise effects. Some of these are beneficial and others are detrimental.

12.1 SUBSTRATE COUPLING

Coupling is capacitive rather than resistive in SOI, therefore lower frequency noise is attenuated. This is illustrated in Figure 12.1, which shows both bulk and some SOI parasitic. This clearly shows that noise coupling generated in the left-hand side device (the aggressor) is connected to the right-hand side (victim) device in bulk material by an electrical resistance determined by the substrate resistivity. In the SOI device, the coupling is through a capacitance from the aggressor device to substrate, then another capacitor from substrate to victim, a much higher attenuation than with the resistive path.

In terms of equivalent circuit, Figure 12.1 can be represented in several ways, depending upon how or whether the substrate is connected to ground, but a typical integrated circuit with a resistive coupling to ground would look like Figure 12.2. As can be seen from the diagram, the noise level is substantially lower with capacitively coupled noise, and a single-sided noise spike is also converted from single-sided to bidirectional.

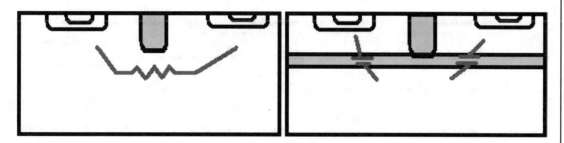

FIGURE 12.1: Demonstration of the coupling between aggressor and victim devices in bulk (left) and SOI (right).

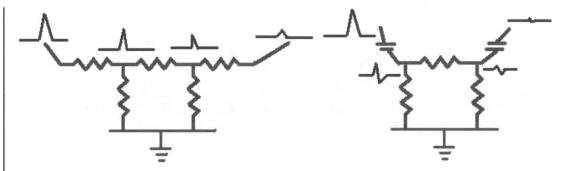

FIGURE 12.2: Equivalent circuit of the coupling between aggressor and victim devices in bulk (left) and SOI (right).

12.2 SUBSTRATE CAPACITANCE/SUPPLY CAPACITANCE

The $V_{dd}-V_{ss}$ capacitance associated with the Nwell of PMOS devices is virtually absent in SOI, being replaced by a very low capacitance SOI capacitor. This is a disadvantage for SOI and means that there is the potential for very much higher supply noise to propagate through the supply line (Figure

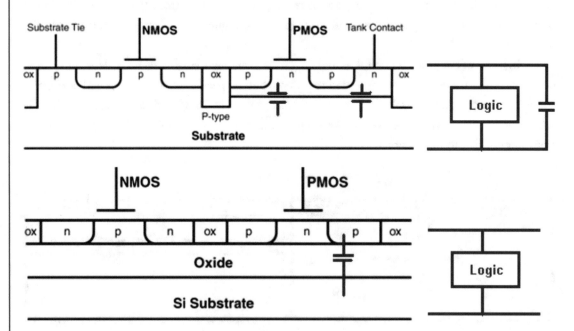

FIGURE 12.3: Cross-sectional and schematic diagrams showing the effect of capacitance due to the Nwell in bulk silicon (upper) as compared to the relative lack of capacitance from SOI (lower).

12.3). To compensate for the lack of chip supply to ground capacitance (and resultant increase in supply noise), it is possible to add junction or gate capacitance in any unused chip areas, or metal capacitance in zones where there is a low density of interconnect.

The capacitance of Nwell to ground junctions in bulk material is generally useful, but it can be an issue in circuits that are switched frequently, as every time the supply is turned on or off, the capacitance associated with the Nwell in substrate has to charge or discharge. In the case of some low-power applications, this may occur many times per second and can result in a significant average current consumption. In such cases, the reduced capacitance and power of SOI can be more advantageous than the increase in noise.

12.3   RADIATION EFFECTS

In addition to improved radiation hardness, the feature sizes of CMOS SOI transistors are readily scalable to well below 0.1 μm and appear, in principle, to be scalable below the limits of bulk CMOS bulk transistors. Scaling of CMOS SOI transistors increases the future potentials of CMOS SOI technology applications not only in radiation hardening but also in general IC processing and manufacturing.

An important advantage of SOI is improvement in SER mainly because of its long history in radiation-hardened applications and the presence of buried oxide. Alpha particles from radioactive elements in packaging are known to induce soft errors and impose design constraints in 6-T planar SRAM cells.

In conventional bulk CMOS, α-generated charges are collected mainly by the funneling effect when particles collide with the drain diffusion layer. This is not significant in SOI MOSFETs due to the presence of buried oxide. Charge collection can only occur in SOI MOSFETs when an α-particle interacts with the channel region. The amount of α-generated charges in SOI MOSFET is less than bulk. The total charge collected at the cell storage node is significantly higher than α-generated charges due to the parasitic bipolar effect. Alpha-induced bipolar current flows over extended time periods. Although one would expect significant SER improvement in SOI, it has also been reported otherwise.

Although use of body contacts reduces parasitic bipolar leakage, density requirements may not permit such an option. The curve for SER (Figure 12.4) of SOI SRAMs behaves differently from bulk SRAM as the gate length is scaled. For bulk SRAM, the scaling of the cell and supply voltage reduces QCRIT, and the SER rises almost exponentially. For SOI SRAM, scaling the supply voltage reduces the parasitic bipolar effect since the body, which acts as the base of the parasitic, will be at a lower potential. The reduced parasitic bipolar effect compensates for the reduction in the QCRIT of SOI SRAM, and the SER remains relatively flat with technology scaling. The noise currents for SOI and bulk behave differently over time.

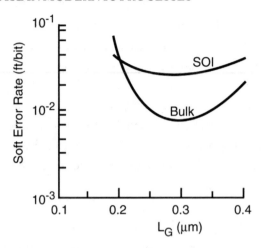

FIGURE 12.4: The 6-T SRAM cell SERs in SOI and bulk SRAM as a function of gate length [25].

The cell size and circuit performance of a SOI SRAM can be further improved by using a cell layout with abutted n+ and p+ drain regions. The n+ and p+ drains of the inverter output node and the source/drain region of the NMOS access transistor are connected by abutting Ti-silicided n+ and p+ regions. This removes layout constraint of well spacing in the bulk CMOS technology and allows a single contact layout for the cross-coupled inverters in the SOI SRAM 6-T cell.

12.4   SOI COMPONENT NOISE

Impact ionization effects are caused when the electrons from the source region (for an NMOS) are accelerated to the point where ionization can occur. This is similar for bulk and SOI and is demonstrated for PDSOI in Figure 12.5.

The act of an electron or hole being captured or released by an oxide trap causes noise. This will be more or less significant depending upon the trap depth within the oxide, its activation energy, and the temperature of operation of the circuit.

Static noise in PDSOI is higher than in bulk due to additional noise being induced through the floating body effects and associated leakage current variability. In PDSOI, there is a parasitic bipolar which can also result in noise. Body potential variability and thus the threshold voltage of PDSOI devices also affect noise. Transistor level static noise analysis has been shown to be extendable to PDSOI technology through the addition of floating body-induced threshold voltage variation and incorporation of parasitic bipolar leakage.

To establish an initial condition for static noise analysis in PDSOI, a choice must be made between minimum and maximum body voltages for each transistor. Typically for memory and digi-

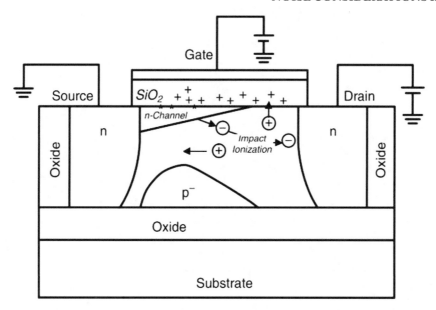

FIGURE 12.5: Electrons from the source and channel region can gain enough energy as they accelerate across to the drain such that an impact causes an electron hole pair to be generated. The hole tends to migrate to the gate oxide, where it can be trapped, and become a source of noise.

tal cells, this might result in a worst case condition of strong noise-inducing devices and weak holding devices. Although noise is normally a problem, one analog device has made beneficial use of noise. A single-stage operational transconductance amplifier in SOI has been used as a generator to produce a high level of noise at its output created by the SOI transistors to produce a random number generator.

References

[1] A. Marshall, "A universal DC to logic performance correlation," International Test Conference, Santa Clara, CA, September 18, 2007.

[2] K. von Arnim, C. Pacha, K. Hofmann, T. Schulz, K. Schrufer, and J. Berthold, "An effective switching current methodology to predict the performance of complex digital circuits," IEDM, December 11, 2007.

[3] K. Agarwal and S. Nassif, "Statistical analysis of SRAM cell stability," DAC, San Francisco, CA, July 24–28, 2006, pp. 57–62. doi:10.1145/1146909.1146928

[4] K. Kang, M. A. Alam, and K. Roy, "Characterization of NBTI induced temporal performance degradation in nano-scale SRAM array using IDDQ," International Test Conference (ITC), 2007. doi:10.1109/TEST.2007.4437590

[5] B. Cheng et al., "Evaluation of intrinsic parameter fluctuations on 45, 32 and 22 nm technology node LP N-MOSFETs," 38th European Solid-State Device Research Conference (ESSDERC), September 15–19, 2008, pp. 47–50.

[6] B. Cheng et al., "Evaluation of statistical variability in 32 and 22 nm technology generation LSTP MOSFETs," 38th European Solid-State Device Research Conference (ESSDERC), 2008.

[7] A. Marshall and S. Natarajan, *SOI Design: Analog, Memory and Digital Techniques*, Springer, 2001, ISBN 0792376404.

[8] A. Marshall, "Thermal coupling of matched SOI device bodies," United States Patent 7,397,085, July 8, 2008.

[9] *Marshall et al., in patent 5432741.

[10] L. Bristol, "Offset balancing method and apparatus for a DC amplifier," United States Patent 4,495,470, January 22, 1985.

[11] J. Devore, A. Marshall, T. McCoy, "Power IC design for testability," 1995 IEEE International Symposium on Circuits and Systems (ISCAS), vol. 2, April 28–May 3, 1995, pp. 1496–99. doi:10.1109/ISCAS.1995.521418

[12] C. Alexander et al., "Random-dopant-induced drain current variation in nano-MOSFETs," *IEEE Trans. Electron. Devices*, vol. 55, no. 11, November 2008, pp. 3251–8.

[13] F. N. Hooge, "1/f Noise is no surface effect," *Phys. Lett.*, vol. 29A, no. 3, April 1969, pp. 139–94.

[14] A. L. McWhorter, "1/f noise and germanium surface properties," Semiconductor Surface Physics, University of Pennsylvania Press, 1957, pp. 207–28.

[15] X. Xi et al., "BSIM4.5.0 MOSFET model—User's manual," Department of Electrical Engineering and Computer Sciences, University of California, Berkeley, CA 94720, 2004.

[16] J. McNeill, "Jitter in ring oscillators," ISCAS 1994, pp. 201–4. doi:10.1109/ISCAS.1994.409561

[17] A. Hajimiri et al., "Jitter and phase noise in ring oscillators," *IEEE J. Solid State Circuits*, vol. 34, no. 6, June 1999, pp. 790–804. doi:10.1109/4.766813

[18] D. Allstot, "Low noise digital logic techniques," *Proc. 4th Annual ASIC Conf.*, pp. pT13-1.1/1.2, 1991. doi:10.1109/ASIC.1991.242866

[19] F. Mendoza-Hernandez, "A noise tolerant technique for submicron dynamic digital circuits," *Rev. Mex. Fis.*, vol. 53, 2007, pp. 72–82.

[20] O. Milter and A. Kolodny, "Crosstalk noise reduction in synthesized digital logic circuits," *IEEE Trans. VLSI Syst.*, vol. 11, no. 6, December 2003, pp. 1153–8. doi:10.1109/TVLSI.2003.817551

[21] A. Marshall et al., "Standard cell power IC design," Custom Integrated Circuits Conference 1998, *Proc. IEEE*, May 11–14, 1998, pp. 329–32. doi:10.1109/CICC.1998.694993

[22] T. Yin et al., "Noise analysis and simulation of chopper amplifier," APCCAS, 2006, pp. 168–70.

[23] http://en.wikipedia.org/wiki/Chopper_(electronics)#Chopper_amplifiers.

[24] http://www.national.com/onlineseminar/2004/chopperamps/Achieving%20Precision.pdf.

[25] C. T. Chuang et al., "SOI for digital CMOS VLSI: Design considerations and advances," *IEEE Proc.*, vol. 86, no. 4, April 1998.

Index

Biography

Andrew Marshall has more than 25 years of experience in the semiconductor industry in a variety of areas including material characterization, process and device development, mixed signal integrated circuit design, and circuit simulation. Dr. Marshall has 45 issued patents and 60 published papers. He is co-author of the book *SOI Design: Analog, Memory and Digital Techniques* and is a Fellow of the Institute of Physics and a Senior Member of the IEEE.

Printed in the United States
by Baker & Taylor Publisher Services